Cotton - New Farming Practices and Products

Edited by Ibrokhim Y. Abdurakhmonov

Published in London, United Kingdom

Cotton - New Farming Practices and Products
http://dx.doi.org/10.5772/intechopen.1008137
Edited by Ibrokhim Y. Abdurakhmonov

Contributors
Dalia Gallico, Feng Shi, Ghulam Sarwar, Honghai Luo, Ibrokhim Y. Abdurakhmonov, Jeyajothi Raman, Jianping Cui, Liwen Tian, Marimuthu Subramani, Muazzama Batool, Muhammad Tauseef, Sivakumar Kaliyannan, Sureshkumar Rajalingam, Vignesh Natarajan

First published in London, United Kingdom, 2026 by IntechOpen

IntechOpen is the global imprint of INTECHOPEN LIMITED, registered in England and Wales, registration number: 11086078, 167-169 Great Portland Street, London, W1W 5PF, United Kingdom

For EU product safety concerns: IN TECH d.o.o., Prolaz Marije Krucifikse Kozulić 3, 51000 Rijeka, Croatia, info@intechopen.com or visit our website at intechopen.com.

British Library Cataloguing-in-Publication Data
A catalogue record for this book is available from the British Library

Cotton - New Farming Practices and Products
Edited by Ibrokhim Y. Abdurakhmonov
p. cm.
Print ISBN 978-1-83635-493-2
Online ISBN 978-1-83635-492-5
eBook (PDF) ISBN 978-1-83635-494-9

Meet the editor

Ibrokhim Y. Abdurakhmonov received his BS in Biotechnology (1997) from the National University of Uzbekistan, an MS in Plant Breeding (2001) from Texas A&M University, and a Ph.D. in Molecular Genetics (2002) and a D. Sc. in Genetics (2009) from the Academy of Sciences of Uzbekistan. He founded the Center of Genomics and Bioinformatics of Uzbekistan in 2012. He received the 2010 TWAS Prize and the ICAC Cotton Researcher of the Year Award 2013 for his outstanding contributions to cotton genomics and biotechnology. Dr. Abdurakhmonov was elected as a fellow of The World Academy of Sciences (TWAS) in 2014 and a member of the Academy of Sciences of Uzbekistan in 2017. In 2023, he was awarded the UNESCO-Guinea Equatorial Life Science Award for his contributions to the development of molecular markers and disease resistance in cotton. He received the 8th VCANBIO Award for Biosciences and Medicine in 2023.

Contents

Preface

Cotton, frequently referred to as "white gold", constitutes an indispensable component of the global agricultural economy. It functions as a crucial raw material for the textile industry and represents a primary source of income for millions of farmers worldwide. Over the years, cotton research and production have undergone significant transformations driven by emerging technologies, environmental challenges, market trends, and evolving policies.

This new book, *Cotton – New Farming Practices and Products*, is a comprehensive guide that covers the landscape of cotton cultivation, breeding, biotechnology, new cotton farming practices, pest management, and sustainability efforts that are helping boost production and improve its use. It presents recent developments and innovations shaping the cotton industry while thoughtfully addressing ongoing challenges, including climate change, resource shortages, and the importance of environmentally sustainable farming practices.

Our goal with this new volume is to share valuable insights into cotton farming, environmental issues, scientific breakthroughs, and the global cotton market, particularly as we confront the challenges of climate change and the impact of technological advancements. By highlighting the latest progress, this volume aims to inspire and guide future efforts in the cotton world, supporting informed and confident decisions by scientists, policymakers, and industry leaders.

I sincerely hope that this publication serves as a helpful and valuable resource for all those engaged in the cotton sector, from the fields where cotton is planted to the textiles used in our garments, and that it encourages ongoing efforts towards a robust and sustainable cotton industry.

The chapters in this book are structured to examine recent advancements in critical fields such as modern biotechnology and breeding, intelligent technology in irrigation and precision agriculture, enhancing productivity through innovative agricultural methods and planting techniques, and the development of the cotton industry into fashion and textiles, including its evolution and value addition over time.

By examining novel scientific concepts and providing a comprehensive perspective on the significance of cotton to our economy and society, I am confident that this volume will serve as a valuable resource for students, professionals, researchers, and all individuals interested in the cotton industry, particularly in new cotton farming practices, textile production, biotechnology, and environmental sustainability.

A heartfelt thank you to all the authors of the chapters for their hard work and valuable contributions. Your insights have enriched this book and will undoubtedly benefit the cotton community. I also want to thank the IntechOpen team, Ms. Elvira Baumgartner, and the reviewers for their invaluable support and guidance throughout this project.

Ibrokhim Y. Abdurakhmonov
Center of Genomics and Bioinformatics,
Academy of Sciences of Uzbekistan,
Tashkent, Uzbekistan

Section 1

Introduction

Chapter 1

Introductory Chapter: Brief Overview of the Latest Updates on Cotton Research and Production

Ibrokhim Y. Abdurakhmonov

1. Introduction

The Gossypium genus includes roughly 50 species, divided into eight diploid genomic groups and a single, monophyletic tetraploid cluster. It is the world's most important natural fiber crop [1, 2]. Furthermore, cottonseed, a byproduct of cotton production, is a major source of edible oil in various countries and has the potential to be a critical source of protein for humans [3]. Given the global importance of this cash crop, it is vital to conduct a timely review of the current state of cotton research and production to forecast future development trends. Several cotton books [4–7] were published on the IntechOpen open-access publication platform [2]. All of these books discussed the most recent advances in cotton research and reported on the periodic accomplishments in cotton science and farming. However, there is a need for continuous periodic updates to address the key challenges and constraints that cotton production faces globally, due to the introduction of new technologies and methods. Several novel approaches are being employed to enhance cotton growth and development, with a focus on increasing production, improving quality, and enhancing environmental sustainability. These advancements cover a wide range of topics, including biotechnology, agricultural precision, and more efficient irrigation methods.

This volume highlights the latest advancements and updates in key areas like breeding, genomics, and digital innovations in cotton farming. It covers exciting technologies such as precision farming, smart and sustainable irrigation, high-density planting, and plastic mulching to boost cotton yields in Xinjiang. The chapters also explore new directions for cotton textiles, value-added chains, and innovative uses of cotton waste for composites. Besides showcasing these detailed advancements and the work of an international group of cotton scientists, I have also included a brief overview of some of the most recent research and game-changing farming practices from the past few years.

2. The research efforts in the last 3 years

To get the most recent updates on cotton research efforts, the PubMed database was searched using the keyword "Gossypium." According to our results, 10,074 publications were retrieved between 1908 and June 2025, representing 1014 more articles in the last 3 years (**Figure 1**). To categorize the PubMed search results, I divided them

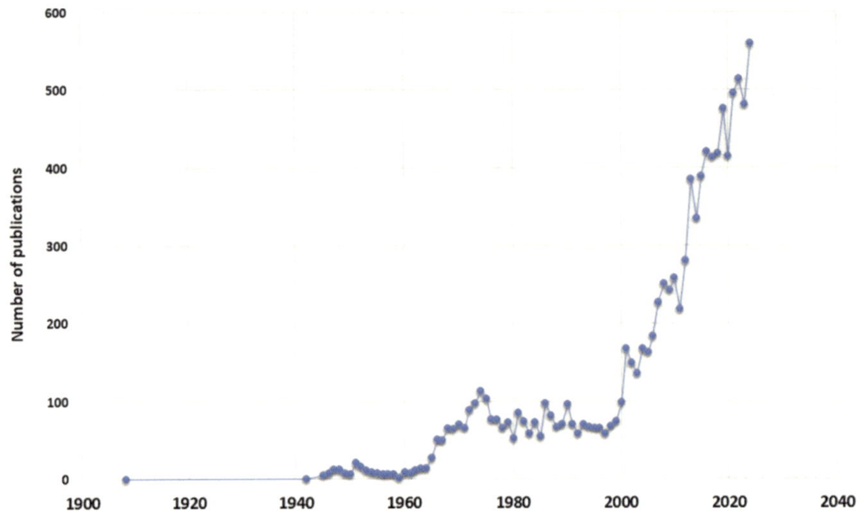

Figure 1.
PubMed [8] identified scientific papers with the keyword "Gossypium" by June 2025.

into suitable groups based on research trends, topics, and applications (**Figure 2**; **Table 1**). Many publications have focused on high-quality genome assemblies and pangenome sequencing, CRISPR/Cas9 gene editing for trait improvement, genome wide association studies (GWAS) and AI-driven breeding, epigenetic and RNAi-based approaches, speed breeding and haploid induction, synthetic biology and microbial symbiosis, and the development of commercialized varieties.

Furthermore, I focused on research efforts conducted over the past few years (2023–2025) to learn about new trends and innovative techniques in cotton. PubMed [8] showed 1283 publications published during this period. Cotton molecular research is a top priority and a simple solution to the challenges. In recent years, there has been an increase in research articles on cotton genome sequencing, gene editing

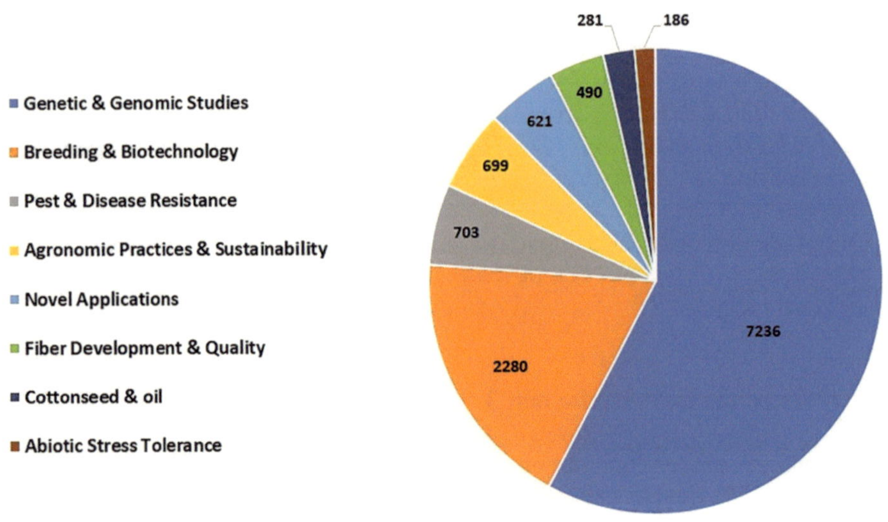

Figure 2.
Research directions of the scientific publications.

No	Research directions of the scientific publications	From	No. of publications
1	**Genetic and genomic studies**		
	Cotton genome sequencing	1972	3238
	Genetic diversity and evolution of Gossypium species	1989	262
	Comparative genomics (diploid vs. tetraploid cotton)	2004	64
	Molecular markers and QTL mapping	1989	3672
2	**Breeding and biotechnology**		
	Cotton breeding for improved traits (yield, fiber quality, drought resistance)	1975	448
	Transgenic cotton (e.g., Bt cotton for pest resistance)	1987	1813
	CRISPR/Cas9 and gene-editing applications	2017	19
	Hybrid cotton development	1970	738
3	**Pest and disease resistance**		
	Resistance to bollworms, aphids, and whiteflies	2013	5
	Bacterial disease resistance	1958	291
	Fungal disease resistance (Fusarium wilt, Verticillium wilt)	1980	223
	Nematode resistance in cotton	1970	184
4	**Abiotic stress tolerance**		
	Drought and salinity tolerance mechanisms	2006	49
	Heat stress adaptation	1988	76
	Heavy metal tolerance (e.g., cadmium, arsenic)	1980	61
5	**Fiber development and quality**		
	Cellulose biosynthesis in cotton fibers	1952	349
	Fiber strength, length, and micronaire studies	1964	1079
	Transcriptomics/proteomics of fiber development stages	2002	141
6	**Agronomic practices and sustainability**		
	Fertilization and nutrient use efficiency	2004	43
	Water-use efficiency in cotton farming	2003	69
	Organic cotton production	1964	577
	Impact of climate change on cotton cultivation	2018	10
7	**Cottonseed and byproducts**		
	Cottonseed oil composition and uses	1951	219
	Gossypol toxicity and reduction strategies	1978	48
	Utilization of cotton byproducts (e.g., livestock feed, biofuels)	1977	14
8	**Economic and social studies**		
	Global cotton trade	1986	30
	Socioeconomic impact of cotton farming	1973	4
	Policy and regulatory issues (e.g., Genetically modified organism (GMO) cotton adoption)	2013	2
9	**Historical and taxonomic research**		
	Evolutionary history of Gossypium species	1974	51
	Domestication and origin of cultivated cotton	1990	18

No	Research directions of the scientific publications	From	No. of publications
10	**Novel applications**		
	Cotton in biomaterials (e.g., wound dressings, textiles)	1954	617
	Nanocellulose derived from cotton fibers	2012	4

Table 1.
The main directions, the initial published years, and the number of publications.

(e.g., CRISPR), and trait-associated markers (239 publications). Cotton research is also heavily focused on breeding and agronomy. In recent years, 76 publications have been published that focus on hybrid varieties, drought-resistant strains, and yield improvement. 278 papers on pest management explained cotton bollworm resistance. Researchers also published 159 publications on disease management, which included fungal diseases (such as *Fusarium*) and biocontrol strategies. Furthermore, 257 publications focused on enhancing the strength and length of cotton fibers. Researchers used biotechnology to publish 607 publications. Surprisingly, 1211 articles focused on environmental impact, including water usage, pesticide reduction, and sustainable farming practices. It could explain global warming and water limitation.

Cotton genomics and breeding have witnessed several significant developments over the last 3 years, driven by advances in sequencing technologies, gene editing, and computational biology. During this time, high-quality genome assemblies and pangenomes were deployed on both tetraploid [9] and diploid cotton [10]. Improved genome assemblies for *Gossypium hirsutum* (upland cotton) and *Gossypium barbadense* (Pima cotton) were published, enabling more accurate trait mapping [9]. In addition, many cotton varieties were sequenced to generate pangenomes, which capture genetic diversity and reveal novel genes associated with fiber quality, stress resistance, and disease tolerance [11–13]. Furthermore, scientists performed telomere-to-telomere (T2T) gapless and high-quality genome sequencing of cotton to resolve complicated repetitive areas, improve trait mapping, and accelerate breeding for drought tolerance and fiber quality [14–16].

The CRISPR/Cas9 gene-editing techniques are being widely utilized to improve important agronomic traits in cotton. For example, GhPGF (resulted in a glandless phenotype in stems and seeds) and GhRCD1 (involved in cotton callus proliferation) were modified using CRISPR/Cas9 to produce glandless stems and seeds while also increasing callus proliferation [17]. Furthermore, a thorough analysis of the Calcium-dependent protein kinases (CDPKs) gene family in upland cotton revealed that GhCPKs are highly expressed in many cotton tissues and respond positively to a variety of biotic and abiotic stresses. A method for screening insect resistance genes from a CRISPR-Cas9 mutant library of GhCPKs was developed [18].

Scientists studied GhIRX10, a beta-1,4-xylosyltransferase responsible for producing the xylan backbone. Overexpression of GhIRX10 promotes xylan synthesis, which affects fiber elongation and secondary cell wall thickness. GhMYB102, identified as a direct regulator of GhIRX10 expression, was subsequently verified through extensive validation. Overexpression of GhMYB102 resulted in a phenotype similar to that of OE-GhIRX10, characterized by higher cell wall thickness and shorter fiber length. Overexpression of GhMYB102 increased the expression of cell wall synthesis-related genes such as GhCESA4/7/8, GhIRXs, GhCESAs, GhGUXs, GhTBLs, GhXTHs, and GhXXTs. Consequently, the cellulose and hemicellulose contents in OE-GhMYB102 lines were dramatically elevated. GhMYB102 was also identified as a target gene regulated by GhFSN1 and GhMYB7, with the ability to reciprocally regulate GhFSN1 expression. In summary, scientists propose a regulatory paradigm in which GhMYB102 enhances the expression of GhIRX10 and other cell wall-related genes, thereby influencing fiber quality [19].

3. Cotton production and use

Cotton production supports many people's lives worldwide, particularly in poorer nations, making it an important component of rural economies. Cotton production, despite its importance, faces numerous challenges, including biotic stressors from pests and diseases, as well as abiotic factors such as drought, salinity, and high temperatures [20]. Biotic stresses degrade cotton fiber quantity and quality, leading to increased use of chemical pesticides, which pose environmental and health threats. Cotton agriculture faces significant challenges from both biotic and abiotic factors, including drought, salinity, and high temperatures. The effects of drought are especially severe in dry and semiarid environments, where a lack of water hinders the development and production of cotton [21]. Salt stress, frequently caused by saline water irrigation or high soil salinity, harms seed germination, plant growth, and fiber development [22]. These difficulties require novel techniques to ensure long-term production. Knowing the status of cotton output on an annual schedule is critical. With the keyword search of "cotton production," 9068 publications were retrieved starting in 1912. Over the last 5 years, more than 3000 publications have been published on this issue. All publications are categorized into five directions, shown in **Table 2**.

Agricultural and crop science has become a crucial field, encompassing techniques for growing cotton, soil health, cotton farming, and pest and disease management. Over the last 5 years, environmental and sustainable development issues have also been explored to address water scarcity, pesticide use, and the increase in organic cotton production. Modern breeders are concerned about pesticide exposure among cotton farmers, occupational risks in cotton processing, and child labor in cotton cultivation. Numerous studies have been conducted on these topics. Approximately

1	Directions	Number of publications	Sub-directions
2	Agriculture and crop research	2974	Cotton cultivation techniques
			Soil health and cotton growing
			Pest and disease management
3	Environment and sustainable development issues	204	Water consumption
			Pesticide impacts and alternatives
			Organic cotton production
4	Healthcare and employment impacts	96	Pesticide exposure in cotton farmers
			Occupational hazards in cotton processing
			Child labor in cotton farming
5	Financial and political studies	22	Cotton trade and economic impacts

Table 2.
Categorized direction of cotton production within the last 5 years.

20 articles, including financial and political studies, explained the cotton trade and its economic repercussions.

Cotton diseases pose substantial concerns for production, quality, and profitability, but sustainable management measures offer a feasible solution. Farmers can reduce disease risks and environmental damage by combining resistant varieties, biological controls, soil health practices, and appropriate chemical use. The transition to sustainability is not only ecological but also economic, as it ensures long-term production and resilience in cotton cultivation. However, issues such as restricted access to sustainable inputs and knowledge gaps among smallholder farmers must be addressed collaboratively by governments, researchers, and the private sector. Precision agriculture innovations, disease-resistant breeding, and farmer education will all contribute to enhancing cotton production systems. Ultimately, sustainable disease control in cotton is about striking a balance—protecting crops today without jeopardizing future resource availability. By adopting integrated, science-based approaches, the cotton industry can ensure a healthier, more sustainable future for farmers, consumers, and the environment.

The ICAC 2024 report forecasts U.S. upland cotton production for the 2024 crop to be 14.6 million 480-pound bales (equivalent to 3.2 million metric tons or MMT). This is a 24% increase compared to the previous year but remains below the 5-year average of 15.2 million bales (3.3 MMT). The planted area in 2024/25 totaled 4.4 million hectares, increasing 9% from the previous year. The harvested area is expected to be 3.4 million hectares, implying a 23% abandonment rate, which is lower than the previous year and less than half the 47% record set in 2022/23. The yield per harvested hectare is expected to be 930 kilos, down from 1000 kilograms last year [23].

In Asia, Chinese scientists and breeders established important technologies of cotton production in Xinjiang. Between 2020 and 2024, the average total fiber yield, planting area, and unit yield of cotton in Xinjiang were 5.296 million tons, 2.4644 million hectares, and 2149.72 kg/ha, respectively. The maximum total yield, planting area, and unit yield were 5.686 million tons, 2.5061 million hectares, and 2322.8 kg/ha, respectively. Furthermore, the cotton quality in Xinjiang is good, and the overall mechanization coverage rate has reached 97.0%. Xinjiang's cotton industry succeeds mainly due to regional cotton planting technologies, such as the "early, short, and dense" method, pest and disaster management, water-saving irrigation, and standardization [24]; see also the related chapter in this volume.

In this context, it is noteworthy to mention that Uzbekistan has made a groundbreaking and historic move in its cotton production by adopting and testing new strategies [25]. Specifically, in 2024, Uzbekistan for the first time introduced (1) Bt-cotton cultivars resistant to cotton bollworms with herbicide tolerance and (2) adopted Xinjiang's agronomic practices, such as high-density planting, plastic mulching, and drip irrigation using 66 x 10 cm double-row plant spacing (**Figure 3**) or 76 cm single-row planting schemes, allowing 200–240 plants to be grown per hectare. In 2024, 18 cotton cultivars from China, India, Turkey, and Spain were transferred and planted on 96 thousand hectares across 13 cotton-growing regions to trial and identify the most suitable cotton varieties. Notably, two Bt and herbicide-resistant Chinese cultivars, "Xin Lu-Zhao-52" and "Xin Lu-Zhao-78," were trialed on 72 thousand hectares.

As a result, the average yield in these 96-hectare trials was 26% higher (4.2 tons/hectare) compared to traditional fields (average of 3.1 tons/hectare), justifying the expansion of Bt cultivars and Xinjiang's agronomic practices in Uzbekistan. Therefore, in the 2025 season, Bt cultivars and Xinjiang's agronomic practices were adopted by

Figure 3.
Xinjiang's agronomic practices in Uzbekistan. (A) 66x10 cm double-row plant sowing under plastic mulching and drip irrigation system, (B) seedling emergence, (C–E) plant growth, and (D) fully ripened field.

32% (281,000 hectares) of the cotton-growing area, which is favored by local farmers over traditional farming methods. Many experienced farmers who fully implemented and followed Xinjiang's agronomic practices (across 6000 hectares in 2024) achieved outstanding yields of 6–7.5 tons/ha, demonstrating the potential for yield boosting in Uzbekistan. The application of 66x10 cm double-row high-density planting (200–240 plants/ha) with non-Bt traditional Uzbek cultivars has also demonstrated a high yield of over 4.5 tons compared to traditionally practiced 90x15 cm single-row planting, which yields only 90–100 thousand plants per hectare. All of these initial achievements encouraged Uzbek farmers to adopt these new farming practices in subsequent seasons. To address the late maturity characteristics and coarse fiber quality of Chinese cultivars compared to Uzbekistan cotton cultivars, the development of Bt and herbicide-tolerant Uzbek cultivars using traditional Uzbek cultivars and Chinese cotton germplasm *via* backcross breeding is ongoing (not published).

4. Soil health and cotton growing

Soil is a valuable and nonrenewable resource for agroecosystems, and its long-term health is becoming a global concern. Current agricultural practices, which involve excessive pesticide and fertilizer use, monoculture, and intensive tillage, have led to widespread soil degradation, highlighting the need for innovative strategies to restore and sustain soil health and functionality [26]. A healthy soil functions as a complex and balanced ecosystem with good structure, an ideal functional state, and a high buffering capacity that maintains a dynamic equilibrium among all its components. There is an urgent need to develop new methods for the long-term management of soil quality in depleted soils to boost agricultural output and ensure food security as the global population continues to grow [26, 27]. The widespread presence of microplastic and nanoplastic pollutants (MNPs) in the environment has raised concerns about their potentially harmful effects on soil health, given their prevalence and durability. However, the interactions of MNPs with soil components, microbial communities, plants, and wildlife—elements that likely have the greatest impact on nutrient availability and food security—remain poorly understood [26]. Several innovative approaches to enhance soil health have been implemented in recent years.

5. Organic cotton production

Organic agriculture promotes biodiversity, soil health, and sustainable ecological management. Farmers emphasize the importance of renewable resources and conserving soil and water to protect the environment for future generations. Organic meat, poultry, eggs, and dairy come from animals not given antibiotics or growth hormones [28]. Organic cotton is grown using natural fertilizers, compost, and biological pesticides, avoiding synthetic fertilizers and GMOs. Organic cotton farming has a low environmental impact, which increases biodiversity and maintains soil nutrients [29]. Organic cotton cultivation depends on habitat management and ecological engineering principles to restore soil health and produce a healthy crop that is less susceptible to insect pests and diseases.

Organic cotton prohibits the use of genetically modified seeds, artificial fertilizers, synthetic insecticides, and chemical plant growth regulators [30]. In 2016, organic cotton production accounted for 0.4% of worldwide output and 0.1% of total area. In 2016, there were approximately 8303 certified organic centers, and around 50 to 60 brands sold organic cotton worldwide. Global organic cotton output climbed from 24,000 tons in 2004 to 240,000 tons in 2009 but has since declined to 108,000 tons in 2016. In 2015–2016, 18 nations manufactured organic cotton. In 2020/21, only eight countries generated an estimated 97% of worldwide organic cotton: India (38%), Turkey (24%), China (10%), Kyrgyzstan (9%), Tanzania (6%), Kazakhstan (4%), Tajikistan (4%), and the United States (2%). The remaining 13 organic cotton-producing countries contributed 3% [31]. In 2022, around 25% of the region's cotton production was certified as sustainable, with India accounting for roughly half of global organic cotton production. Exports of sustainable cotton from the region have increased, with roughly 15% of cotton exports in 2022 coming from sustainably certified sources [32].

In India, breeders are growing organic cotton by installing pheromone traps at 50-meter intervals, with five traps per acre per bug, to aid in monitoring bollworm complexes, including the American bollworm, pink bollworm, spotted bollworm, and tobacco cutworm. Furthermore, installing light traps at a rate of 1–2 traps per acre can aid in monitoring the insect activity of the American bollworm and tobacco cutworm. Adult whiteflies' activities can be observed by placing yellow pan traps and sticky traps at a height of 30 cm above the plant canopy [33]. Bukhonka et al. [34] provided a 15-year history demonstrating the global growth of organic cotton output from 2006 to 2007 to 2020–2021.

6. Conclusion

Cotton science has made significant progress in recent years, thanks to advances in high-quality whole-genome, pangenome, and T2T sequencing, CRISPR editing, and sustainable agriculture. These innovations, from drought-resistant varieties to precision breeding, address global challenges like climate change and food security.

However, transforming laboratory discoveries into scalable, environmentally friendly farming methods remains essential. Going forward, integrating AI, multi-omics, and international collaboration will be vital for ensuring the sustainability and productivity of cotton for future generations of farmers. The transfer of technology, sharing productive cultivars, and adopting new farming techniques, such as Xinjiang's high-density planting and plastic mulching practices, show promise for increasing cotton yields.

Author details

Ibrokhim Y. Abdurakhmonov
Center of Genomics and Bioinformatics, Academy of Sciences of Uzbekistan, Republic of Uzbekistan

*Address all correspondence to: i.y.abdurakhmonov@gmail.com

References

[1] Reddy U, Nimmakayala P, Abburi V, et al. Genome-wide divergence, haplotype distribution and population demographic histories for Gossypium hirsutum and Gossypium barbadense as revealed by genome-anchored SNPs. Scientific Reports. 2017;7:41285. DOI: 10.1038/srep41285

[2] Abdurakhmonov IY. Introductory chapter: Global cotton research development trends for the past five years—Key directions. In: Cotton Research. London: IntechOpen; 2022. DOI: 10.5772/intechopen.105748

[3] Wu M, Pei W, Wedegaertner T, Zhang J, Yu J. Genetics, breeding and genetic engineering to improve cottonseed oil and protein: A review. Frontiers in Plant Science. 2022;**13**:864850. DOI: 10.3389/fpls.2022.864850

[4] Abdurakhmonov IY. World Cotton Germplasm Resources. London: IntechOpen; 2014. p. 322. DOI: 10.5772/56978

[5] Abdurakhmonov IY. Cotton Research. London: IntechOpen; 2016. p. 310. DOI: 10.5772/61914

[6] Mehboob-Ur-Rahman M, Zafar Y. Past, Present and Future Trends in Cotton Breeding. London: IntechOpen; 2018. p. 184. DOI: 10.5772/intechopen.69672

[7] Ansari M. Advances in Cotton Research. London: IntechOpen; 2020. p. 84. DOI: 10.5772/intechopen.77828

[8] National Center for Biotechnology Information (NCBI). [Internet]. PubMed Search Results for "Gossypium" (1908-2025). 2025. Available from: https://pubmed.ncbi.nlm.nih.gov/?term=%27gossypium%27&filter=years.1908-2025 [Accessed: July 3, 2025]

[9] Chang X, He X, Li J, Liu Z, Pi R, Luo X, et al. High-quality *Gossypium hirsutum* and *Gossypium barbadense* genome assemblies reveal the landscape and evolution of centromeres. Plant Communications. 2024;**5**(2):100722. DOI: 10.1016/j.xplc.2023.100722

[10] Fu C, Wang N, Meng Q, Xu Z, Le Y, Zhang X, et al. *Gossypium latifolium* genome reveals the genetic basis of domestication of upland cotton from semi-wild races to cultivars. Crop Journal. 2025;**13**(3):929-941. DOI: 10.1016/j.cj.2025.02.014

[11] Meng Q, Xie P, Xu Z, Tang J, Hui L, Gu J, et al. Pangenome analysis reveals yield- and fiber-related diversity and interspecific gene flow in *Gossypium barbadense* L. Nature Communications. 2025;**16**(1):4995. DOI: 10.1038/s41467-025-60254-x

[12] Li J, Yuan D, Wang P, et al. Cotton pan-genome retrieves the lost sequences and genes during domestication and selection. Genome Biology. 2021;**22**:119. DOI: 10.1186/s13059-021-02351-w

[13] Zhao Z, Zhu Z, Jiao Y, et al. Pan-genome analysis of GT64 gene family and expression response to Verticillium wilt in cotton. BMC Plant Biology. 2024;**24**:893. DOI: 10.1186/s12870-024-05584-6

[14] Yang Z, Li F. From gaps to insights: Telomere-to-telomere cotton genome deciphers centromere dynamics after polyploidization. Science China.

Life Sciences. 2025;**68**:2496-2497. DOI: 10.1007/s11427-025-2918-2

[15] Huang G, Bao Z, Feng L, et al. A telomere-to-telomere cotton genome assembly reveals centromere evolution and a Mutator transposon-linked module regulating embryo development. Nature Genetics. 2024;**56**:1953-1963. DOI: 10.1038/s41588-024-01877-6

[16] Hu G, Wang Z, Tian Z, et al. A telomere-to-telomere genome assembly of cotton provides insights into centromere evolution and short-season adaptation. Nature Genetics. 2025;**57**:1031-1043. DOI: 10.1038/s41588-025-02130-4

[17] Ge X, Xu J, Yang Z, Yang X, Wang Y, Chen Y, et al. Efficient genotype-independent cotton genetic transformation and genome editing. Journal of Integrative Plant Biology. 2023;**65**(4):907-917. DOI: 10.1111/jipb.13427

[18] Wang F, Liang S, Wang G, Hu T, Fu C, Wang Q, et al. CRISPR-Cas9-mediated construction of a cotton CDPK mutant library for identification of insect-resistance genes. Plant Communications. 2024;**5**(11):101047. DOI: 10.1016/j.xplc.2024.101047

[19] Wu A, Shen T, Lu J, Fu X, Yang M, Sun M, et al. GhMYB102 affects cotton fibre elongation and secondary wall thickening by regulating GhIRX10 in cotton. Plant Biotechnology Journal. 2025;**23**(4):1329-1344. DOI: 10.1111/pbi.14588

[20] Thangaraj A, Kaul R, Sharda S, Kaul T. Revolutionizing cotton cultivation: A comprehensive review of genome editing technologies and their impact on breeding and production. Biochemical and Biophysical Research Communications. 2025;**742**:151084. DOI: 10.1016/j.bbrc.2024.151084

[21] Meshram JH, Singh SB, Raghavendra KP, Waghmare VN. Drought. stress tolerance in cotton: Progress and perspectives. In: Climate Change and Crop Stress. Academic press, Elsevier Inc; 2022. pp. 135-169

[22] Maryum Z, Luqman T, Nadeem S, Khan SMUD, Wang B, Ditta A, et al. An overview of salinity stress, mechanism of salinity tolerance and strategies for its management in cotton. Frontiers in Plant Science. 2022;**13**:907937

[23] International Cotton Advisory Committee (ICAC). Meetings & Events. Available from: https://www.icac.org/Meetings/ [Accessed: 3 October 2024]

[24] Tian L, Shi F, Cui J, Luo H. Analysis of Cotton Production Status and Key Supporting Technologies in Xinjiang. InTechOpen; 2025. Available from:. DOI: 10.5772/intechopen.1011390

[25] Abdurakhmonov IY. Sino-Uzbek agro-cooperation fruitful. China Daily. 2025. Available from: https://www.chinadaily.com.cn/a/202506/17/WS6850a5d5a310a04af22c698d.html [Accessed: September 24, 2025]

[26] Țopa D-C, Căpșună S, Calistru A-E, Ailincăi C. Sustainable practices for enhancing soil health and crop quality in modern agriculture: A review. Agriculture. 2025;**15**(9):998. DOI: 10.3390/agriculture15090998

[27] De Corato U, Viola E, Keswani C, Minkina T. Impact of the sustainable agricultural practices for governing soil health from the perspective of a rising Agri-based circular bioeconomy. Applied Soil Ecology. 2024;**194**:105199. DOI: 10.1016/j.apsoil.2023.105199

[28] Lumen Learning. Sustainable Agriculture [Internet] 2023.

Available from: https://courses.lumenlearning.com/suny-monroe-environmentalbiology/chapter/9-5-sustainable-agriculture/ [Accessed: January 8, 2023]

[29] Gallant International. What Is Organic Cotton? [Internet] 2025. Available from: https://www.gallantintl.com/blogs/what-is-organic-cotton/ [Accessed: April 2, 2025]

[30] Beltwide Cotton Conferences, New Orleans, Louisiana, January 9-12, 2007. [Internet] 2007. Available from: https://www.cotton.org/beltwide/proceedings/2005-2022/data/conferences/2007/papers/5849.pdf/ [Accessed: January 12, 2007]

[31] Textile Exchange. Organic Cotton Market Report 2022. [Internet] 2022. Available from: https://textileexchange.org/app/uploads/2022/10/Textile-Exchange_OCMR_2022.pdf/ [Accessed: Oktober, 2022]

[32] Fairtrade International. Cotton. [Internet] 2025. Available from: https://www.fairtrade.net/napp-en/our-products/our-products/Cotton.html/ [Accessed: July 11, 2025]

[33] Confederation of Indian Industry (CII). Sustainable Cotton Cultivation: Practices and Policies in India. [Internet] 2023. Available from: https://caionline.in/uploads/publications/doc/21_22-08-2023.pdf/ [Accessed: August 22, 2023]

[34] Bukhonka N, Pesic M, Nemesa I, et al. Global development trends in organic cotton. In: Proceedings of the VII International Scientific-Practical Conference. Ukraine: Kyiv; 2023. pp. 157-162

Section 2

Cotton Production and New Farming Practices

Chapter 2

Analysis of Cotton Production Status and Key Supporting Technologies in Xinjiang

Liwen Tian, Feng Shi, Jianping Cui and Honghai Luo

Abstract

This article deeply analyzes the current situation and key technologies of cotton production in Xinjiang. From 2020 to 2024, the average total fiber yield, planting area, and unit yield of cotton in Xinjiang reached 5.296 million tons, 2.4644 million hectares, and 2149.72 kg/ha, respectively. Among them, the highest total yield, planting area, and unit yield were 5.686 million tons, 2.5061 million hectares, and 2322.8 kg/ha, respectively. Additionally, the cotton quality of Xinjiang is excellent, and the comprehensive mechanization operation coverage rate has reached 97.0%. The success of Xinjiang's cotton industry is primarily attributed to the development and promotion of a series of cotton planting technologies with distinct regional characteristics and leading technological levels, including identifying the "early, short, and dense" cultivation mode as a technological breakthrough, integrating comprehensive prevention and control of pest, disease, weed, and natural disaster, water-saving through mulched drip irrigation, and standardized management of mechanically harvested cotton. These technologies fully demonstrate the high adaptability of local cotton production techniques to natural conditions. Through these advanced technologies, the unfavorable factors of relatively poor natural ecological conditions in Xinjiang's cotton region have been overcome, and the goal of high and stable cotton yield, high quality, simple, and efficient, has been successfully achieved. Xinjiang cotton not only provides strong support for local economic and social development, but also offers a "China Xinjiang cotton cultivation application plan" for the high quality and sustainable development of global cotton, particularly in cotton production areas with similar natural ecological conditions as Xinjiang.

Keywords: Xinjiang cotton production, "early short and dense" planting mode, integrated management of pests, diseases, weeds and natural disasters, mulched drip irrigation, machine picked

1. Introduction

As one of the world's important cotton-producing countries, China's production has consistently ranked among the top two in the world, which is inseparable from the solid support provided by cotton production in Xinjiang. The Xinjiang cotton region is located in the inland area, spanning 34°20′ N ~ 47°15′ N latitude, consisting of multiple oasis plains separated by the Gobi Desert. The climate belongs to a typical temperate continental climate, with dry and rainless summers and an average annual precipitation of only about 125 millimeters, but abundant light resources and a large temperature difference between day and night; winters are extremely cold and dry, resulting in generally lower pressure for the occurrence and prevention of pests and diseases, and the peak period of alpine snowmelt in spring and summer highly coincides with the critical water demand period for cotton. In addition, cotton fields in this region are characterized by significant human management and control [1, 2]. This region's unique geographical and ecological environment has laid a solid natural foundation for high yield and high quality in the cotton-growing areas of Xinjiang. However, the Xinjiang cotton region faces multiple challenges, including a short frost-free period, relatively limited heat resources, uneven spatial and temporal distribution of light and heat resources, seasonal water shortages, severe soil salinization, poor soil fertility, and weak water and fertilizer retention capacity. These factors led to multiple constraints and challenges to the ecological environment system of the cotton growing areas, seriously affecting its stability and thus threatening the sustainable development of cotton production in Xinjiang [3–8].

To address the above-mentioned challenges, Xinjiang's scientific and technological personnel have jointly tackled the problem and constructed a modern cotton production technology system suitable for the local natural ecological conditions. This system is based on the "early, short, and dense" planting mode, integrating standardized management of key technologies, such as precision seeding and supporting seedling protection, precision water and fertilizer integration under plastic mulch drip irrigation, prevention and control of pests, weeds, diseases, and natural disasters, and integration of machinery and agronomy. These measures vigorously promoted the transformation of Xinjiang's cotton industry toward light, simple, efficient, and green development. As of now, the comprehensive mechanization rate of cotton in Xinjiang has reached 97.0%, with the cotton fields achieving full process mechanical operation coverage from soil preparation, sowing, tillage, water and fertilizer management, chemical topping, defoliation, and ripening to mechanical harvesting. Notably, a small number of core demonstration fields have achieved unmanned management and control of cotton planting by relying on modern intelligent information technologies, such as the Internet of Things (IoT), big data, and artificial intelligence (AI) [9–15].

Due to the systematic and efficient technological development in Xinjiang's cotton region, Xinjiang cotton has led the country in five indicators of total production, yield per unit, commodity export rate, outward transfer volume, and per capita possession for 32 consecutive years since 1993. The total annual Xinjiang cotton production is now stable at over 5.1 million tons, accounting for over 90% of the country's total cotton production. Even compared with other major cotton-producing countries worldwide, Xinjiang's cotton planting area, yield per unit, and total production are still among the top, and the overall cotton quality is excellent. Undoubtedly, Xinjiang has become an important cotton-producing area recognized by China and even the world [16–18].

This article provides an objective and systematic review of the current status of cotton production and its supporting technologies in Xinjiang, aiming to present the breakthrough achievements of Xinjiang's cotton planting technology system to domestic and international peers, and offer replicable technical application solutions for global cotton-producing regions, particularly cotton production areas with ecological conditions similar to Xinjiang. This has important demonstration significance and reference value for promoting the high-quality development of the global cotton industry, while enhancing Xinjiang's recognition and influence in the global cotton industry landscape.

2. Current situation of cotton production in Xinjiang

2.1 Production scale

Statistical data show that from 2020 to 2024, the average annual total production, planting area, and unit yield of cotton in Xinjiang reached 5.296 million tons, 2.4644 million hectares, and 2149.7 kg/ha, respectively. Among them, the total production in 2024 reached a historic high of 5.686 million tons, accounting for 92.2% of China's total production and 22.0% of the global total production, respectively; the planting area reached a peak of 2.5061 million hectares in 2021, accounting for 82.8% of China's total planting area and 7.8% of the global planting area respectively; the yield per hectare reached 2322.8 kg/ha in 2024, which was 1.1 times the national average yield per hectare and 2.9 times the world average yield per hectare. As one of the world's core cotton-producing regions, although Xinjiang's planting area accounts for only 7.8% of the global total, it has ultimately produced more than one-fifth of the world's total cotton production by relying on significant technological and yield per unit advantages (**Table 1**) [19, 20].

2.2 Fiber quality

The fiber types of cotton in Xinjiang exhibit obvious unitary characteristics, mainly reflected in the composition of raw cotton, with upland cotton accounting for over 99%, while island cotton accounts for only about 1%. Among them, upland cotton fibers exhibit a bright white appearance with excellent internal quality indicators; island cotton fiber appearance color is slightly inferior to that of upland cotton, but they are still of the pure white type, and their main fiber internal quality indicators are more outstanding (**Table 2**). Although Xinjiang island cotton has shortcomings in terms of maturity, mercerization, and elasticity indicators compared to the best quality island cotton varieties in Giza strain of Egypt (such as Giza 45 and Giza 87), resulting in not entirely satisfactory quality in dyeing uniformity and color fastness, its fibers possess characteristics of ultra-long and ultra-fine texture, high length uniformity index, and few impurities, which still endowed it with unique advantages in spinning ultra-high count yarns of 120 s or even above 200 s. Especially in the context of Giza 45 and Giza 87 being unable to be imported, Xinjiang island cotton has become the choice for domestic textile enterprises to produce high-end yarn raw materials.

Year	2020	2021	2022	2023	2024	Average
Total production (million tons)	5.161	5.129	5.391	5.112	5.686	5.296
Area (million hectares)	2.5019	2.5061	2.4969	2.3693	2.4479	2.4644
Yield per hectare (kg/ha)	2062.7	2046.4	2158.9	2157.8	2322.8	2149.7

Table 1.
Cotton production in Xinjiang from 2020 to 2024.

Composition of raw cotton	2.5% span length (mm)	Strength (cN/tex)	Uniformity Index (%)	Elongation (%)	Reflectance (%)	Micronaire value
Upland cotton	29.5 ~ 33.1	29.6 ~ 33.8	84.5 ~ 86.5	6.8 ~ 7.9	75 ~ 87	3.9 ~ 4.5
Island cotton	37.2 ~ 40.1	43.8 ~ 47.5	86.7 ~ 89.6	6.8 ~ 8.8	74 ~ 78	3.8 ~ 4.3

Table 2.
Main fiber quality index parameters of upland cotton and island cotton.

Due to Xinjiang producing a large amount of high-quality cotton, it plays an irreplaceable role in ensuring the raw material supply for large cotton textile enterprises in China. Prior to the "Xinjiang Cotton Incident", China's total textile and clothing exports accounted for as high as 33.7% of the world's total textile and clothing exports, of which a considerable proportion of the raw materials for cotton products came from Xinjiang cotton, which fully demonstrates the enormous contribution of Xinjiang's cotton industry to the global textile market [21, 22].

However, currently approximately 10% of cotton in Xinjiang, including upland cotton and island cotton, still relies on manual picking methods. Due to manual harvesting, the problem of "three silk" (heterogeneous, heteromorphic, and heterochromatic fibers) contamination inevitably occurs, which to a certain extent limits the production of bleached yarns and the quality assurance of high-end cotton textiles.

3. Key technologies of cotton production in Xinjiang

3.1 "Early, short, dense" cultivation techniques theory

In response to the climate characteristics of short frost-free period and limited heat in the Xinjiang cotton region, an innovative planting model of "early, short, and dense" with "early" as the core has been proposed. Through a series of technical measures to promote "early" growth, accelerate the cotton growth process, and achieve precise adaptation of cotton plant growth to climatic conditions, to achieve the technical goal of "seedlings in April, buds in May, flowers in June, bolls in July, cotton boll opening in August in northern Xinjiang, and cotton boll opening in September in southern Xinjiang" [12, 23–25]. The specific implementation path is as follows:

1. Early sowing at the appropriate time. Cotton field sowing operation is usually completed before April 20th. Although at this time (especially in early and middle of April), the soil temperature at 5 cm underground is generally lower than the stable temperature threshold of above 14°C or more required for seed germination, studies have found that plastic film mulching can increase the soil temperature in the 5 cm soil layer by 2.5–4.5°C, which effectively meets the suitable temperature conditions required for of cotton seed germination and emergence. Therefore, plastic film covering technology is the most crucial and preferred supporting measure for early sowing operations in Xinjiang cotton fields [26];

2. Planting early maturing varieties. Considering the frost-free periods in southern and northern Xinjiang are 180–220 days and 140–180 days, respectively, which are relatively short, and there are significant differences in climate characteristics between the two cotton regions, it is specified that the full growth period of the main cotton varieties planted in southern and northern Xinjiang should be

124–128 days and 118–123 days, respectively, to ensure that the rate of pre-frost flowering reaches over 85% in normal years;

3. Early intertillage. The intertillage operations are carried out immediately after sowing. This technology effectively raises soil temperature and improves soil aeration by dredging the soil, which can synergistically promote the early growth of cotton seedlings and early development of strong seedlings, while also having the role of removing weeds in the field;

4. Early chemical regulation. Generally, mepiquat chloride is sprayed 1–2 times during the cotton seedling and bud stage, with a dosage of 15.0–37.5 g/ha. This technology can ensure stable and healthy growth of cotton plants during the seedling and bud stage, laying the foundation for building a high-yield population structure in the later growth stages;

5. Early watering. Generally, large-scale drip irrigation of cotton fields starts in early June. However, for sandy cotton fields with poor water and fertilizer retention, due to its drought stress that occurs earlier, the first drip irrigation can be advanced to late May, to meet the cotton growth of water needs in the early stages, and thus ensure the formation of high-yield types and populations [27];

6. Early fertilization. In drip-irrigated cotton fields, the precise water fertilizer integrated management strategy of "one water, one fertilizer" is followed. Specifically, fertilizer can be advanced to early June, or even late May, if one chooses "sowing under insufficient soil moisture", drip irrigation is required to supply soil moisture after sowing, and the corresponding topdressing time will be even earlier. This technology effectively ensures balanced development of cotton plants by following the water and fertilizer requirements at different growth stages, while also improving the utilization efficiency of water and fertilizers [27];

7. Early topping. Topping (removing the apical buds of cotton plants) in southern Xinjiang is generally completed before July 15, while in northern Xinjiang, the operation is usually carried out about 10 days earlier than in southern Xinjiang. Through timely topping, it can effectively inhibit apical dominance, reduce the production of ineffective buds, and then optimize the distribution of photosynthetic products to the cotton bolls to avoid the ineffective consumption of nutrients;

8. Spray defoliant ripening agent at the appropriate time. The defoliant ripening agent spraying operation is usually completed in southern Xinjiang before September 25th, while northern Xinjiang usually completes it about 10 days earlier than southern Xinjiang. By grasping the optimal application period of defoliant ripening agents, the efficacy of the agents can be fully exerted to promote cotton leaf shedding and accelerate boll maturation, and ultimately to ensure a higher cotton mechanical picking rate and quality of cotton;

9. Timely cessation of irrigation and early harvesting. Irrigation is generally stopped at the end of August, with centralized harvesting in October. By timely cutting off the water supply in the later stage of cotton growth, nutrients are promoted to be transported to the bolls, ensuring that boll opening is early, concentrated, and sufficient, thereby improving fiber maturity and quality [25];

10. Early detection and early prevention. By advancing the prevention and control measures of pests, diseases, and weeds to the seedling stage or even before sowing, a comprehensive regional prevention and control system covering cotton fields and their surrounding environments is established. Combined with early warning of pests and diseases during the growth period, the threat to cotton production is minimized to the greatest extent possible, thus ensuring the stability of the cotton field ecosystem and high yield and high quality of cotton [18].

Dwarfing cultivation is another key link of the "early, short, dense" planting model for drip-irrigated cotton fields in Xinjiang. Among them, the height of land cotton plants is controlled at 75–85 cm, the internode length of the main stem is stable at around 6.7 cm, and the length of fruiting branches is maintained at 15–25 cm; the height of island cotton plants is maintained at 90–110 cm, the internode length of the main stem is about 6.0 cm, and the length of fruiting branches is 10–20 cm. Through precise dwarfing management, it can not only ensure good ventilation and light transmission conditions inside the cotton field but also meet the adaptation requirements of mechanized harvesting.

Dwarfing cultivation takes variety selection and systematic collaborative regulation as its core. In terms of variety selection, priority is given to cotton varieties with stable growth vigor, balanced vegetative and reproductive growth, and sensitivity to mepiquat chloride regulators, to ensure a genetic foundation for establishing plant dwarf populations. In terms of system regulation, the strategy of "taking water and fertilizer regulation as the main approach and chemical regulation as the supplementary" is clearly defined, and the details of the regulation plan are shown in sections 3.3 and 3.4. This can effectively prevent the risks of "high (excessive high plant style), large (excessively large canopy) and empty (sparse boll setting)" in the cotton field population [17].

Optimizing plant population density is an indispensable key strategy in the "early, short, and dense" planting mode in Xinjiang cotton region. The harvesting density should be controlled at 165,000–187,500 plants/hm^2 in the southern Xinjiang cotton region, while the harvesting density should be controlled at 180,000–202,500 plants/hm^2 in the northern Xinjiang cotton region, all of which have adopted a wide-narrow row planting pattern of 66 + 10 cm (or its derivatives) with an average row spacing of 38 cm. By establishing a high-density planting structure of "small individuals, large population", the leaf area coefficient of cotton fields can reach a relatively high value earlier, thereby fully utilizing the early light and heat resources in the early stage, and creating favorable conditions for boll setting. This is the core of achieving breakthroughs in yield per unit area. To ensure a higher density of cotton fields, the following supporting measures must be implemented:

1. Variety and density collaborative adaptation technology. Select dwarf varieties with compact plant shape, upward-pointing leaves, and moderate leaf area to ensure good ventilation and light transmission properties in high-density drip-irrigated cotton fields with a density of up to 202,500 plants/hm^2. The density setting follows the principle of dynamic adjustment. Specifically, cotton fields with sandy soil, insufficient light and heat, insufficient water and fertilizer retention capacity, or limited water and fertilizer supply are suitable for high-density planting, on the contrary, the density should be appropriately reduced;

2. High germination rate seed production technology. In response to the problem that the germination rate of conventional qualified seeds is only 80%, which is difficult to meet the needs of seedling preservation in cotton fields with precision sowing, the research and development proposes a high germination rate seed production process of "to be selected seeds → polishing → wind selection → gravity selection → color selection → magnetic selection → coating → packaging". Among them, color selection can effectively remove immature yellow seeds, and magnetic selection uses a combination of 80–200 mesh iron powder and N35–N52 neodymium iron boron magnetic rollers to accurately remove crack seeds. Through the above process, the germination rate of the finished seeds is stably increased to about 90% [28];

3. High-quality precise seeder. Compared with the commonly used air suction precision seeders at home and abroad, the self-developed socket mechanical seeder is selected, which not only significantly reduces the power consumption and maintenance costs, but also realizes higher seeding quality. Data show that the demonstration cotton fields using the socket mechanical seeder have an empty hole rate of ≤0.6%, a single seed qualification rate of ≥99.2%, the misalignment rate close to 0, and the seedling retention rate has stably exceeded 80% [29];

4. System collaborative regulation technology. "Dense" using the same systematic collaborative regulation technology and "short", which can effectively avoid the risk of population canopy closure in cotton fields through the strategy of "taking water and fertilizer regulation as the main approach and chemical regulation as the supplementary".

In summary, the "early, short, and dense" planting mode in Xinjiang cotton region establishes a highly collaborative mutual promotion of symbiosis system. "early" is the core strategy, "short" and "dense" are necessary for "early", while providing guarantee for "early"; "short" creates spatial conditions for "dense (high plant population density)" by shaping compact individual plants; "dense" relying on the advantages of plant population to make up for the limitations of individual plant development, effectively reducing the waste of light, heat, and water resources in the early growth stage of cotton. In response to the "early, short, and dense" planting mode, it is still necessary to adapt to local conditions in production practice, and deeply integrate management measures such as water and fertilizer management, chemical regulation, and plastic film mulching, to fully exert the production potential of this planting mode [24, 25].

3.2 Integrated management of pests, diseases, weeds, and natural disasters

3.2.1 Major pests prevention and control

3.2.1.1 Aphid

Source prevention and control: grasp the early spring prevention and control node, southern Xinjiang at the late March to early April, northern Xinjiang at the mid- to late April, uniformly spray 40% acetamiprid diluted 1500–2000 times on overwintering hosts such as flowers and greenhouse vegetables to reduce the base population of insects from the source.

Physical trapping: After the cotton seedlings emerge, hang yellow boards at the density of 600–750 pieces/hm^2 for trapping.

Chemical and biological prevention and control: when the aphid plant rate in the field reaches 30–50%, the rolled leaf/oil-contaminated plant rate reaches 5–10%, and the population continues to rise, select selective insecticides such as flonicamid, pymetrozine, acetamiprid, imidacloprid, or plant-derived pesticides such as matrine and rotenone for spraying control [30].

3.2.1.2 Spider mite

Field cleaning and source control: removing weeds in and around the field to destroy the habitat and breeding grounds of cotton spider mites, thereby reducing the base population of pests from the source.

Hierarchical prevention and control: Adhere to the principle of "early detection and early prevention", when the mite-infested plant rate is below 15% (in the stage of point piece occurrence), focus chemical control on center mite-infested plants; when the mite-infested plant rate exceeds 15%, whole field chemical control should be carried out. Select acaricides such as etoxazole, abamectin, spiromesifen, and pyridaben for spraying control [30].

3.2.1.3 Bollworm

Green prevention and control: timely carry out autumn plowing, winter irrigation, and field cleaning operations after the autumn harvest, to destroy the overwintering grounds of cotton bollworms and reduce the overwintering base number; plant baiting strips of corn and abutilon around the cotton field to trap cotton bollworms and clean up and eliminate the trapped pests in time; simultaneously installing frequency vibration solar insecticidal lamps and cotton bollworm traps to lure and kill cotton bollworms.

Chemical and biological prevention and control: priority is given to the use of biological pesticides such as cotton bollworm nucleopolyhedrovirus, *Bacillus thuringiensis*, and azadirachtin; high-efficiency and low-toxicity chemical agents such as chlorantraniliprole, lufenuron, indoxacarb, and hexaflumuron should be rationally selected for spraying control.

Seed source prevention and control: strengthening the cultivation and promotion of new varieties resistant to cotton bollworm [30].

3.2.2 Major diseases prevention and control

3.2.2.1 Fusarium wilt and Verticillium wilt

Chemical prevention and control: from seedling stage to bud stage, before or at the initial stage of disease occurrence, pesticides such as *Bacillus subtilis*, amino oligosaccharins, ethyl garlicin, hymexazol, and spinosad should be sprayed or drip irrigation should be used for prevention and control.

Crop rotation system: in response to the plots with serious fusarium wilt and verticillium wilt (disease incidence rate $\geq$ 30%), a crop rotation system should be implemented. Non-parasitic crops such as wheat, maize, tomatoes, sugar beet, melons, etc., should be selected for planting for 2–3 years, planting cotton again.

Seed source prevention and control: strengthening the cultivation and promotion of cotton varieties with high resistance (tolerance) to fusarium wilt and verticillium wilt [30].

3.2.2.2 Rhizoctonia solani and Fusarium moniliforme

Seed coating: in response to the seedling diseases such as *Rhizoctonia solani* and *Fusarium moniliforme*, the seed coating agents such as fludioxonil, metalaxyl-M, and azoxystrobin should be used for prevention and control.

Emergency prevention and control: at the early stage of the disease, especially during low temperature and rainy weather, pesticides such as *Bacillus subtilis*, polymyxin, and hymexazol should be sprayed timely for prevention and control.

Field management: sowing under suitable conditions of soil moisture and temperature; timely intertillage after sowing to improve soil temperature and soil aeration, and enhance the resilience of cotton seedlings [30].

3.2.3 Major weeds prevention and control

Chemical sealing: After the completion of soil preparation operation and before harrowing, 33% pendimethalin EC at a dosage of 2250–2700 mL/hm^2 should be sprayed uniformly on the soil. In response to high incidence field of *Solanum nigrum*, 50% pargyline at a dosage of 1500–2250 g/ha or 42% fludioxonil at a dosage of 1500–2250 g/ha should be selected for soil sealing treatment.

Agronomic measures: during the seedling and bud stage, the intertillage operation should be carried out to remove weeds between the rows. For weeds that cannot be removed by intertillage, manually remove them. In addition, before harvesting cotton, it is necessary to thoroughly remove weeds in the field to prevent their seeds from maturing and falling into the cotton field, which would increase the difficulty of weed control in the following year and also prevents weed contamination of cotton during machine harvesting [17].

Seed source prevention and control: accelerating the cultivation and promotion of cotton varieties with herbicide-resistant property.

3.2.4 Major natural disasters prevention and control

3.2.4.1 Low-temperature frost disaster

Preventive measures: adjust the sowing period according to the local climatic conditions to avoid the frequent occurrence of low-temperature frost disasters.

Post-disaster management: if low-temperature freezing damage cannot be avoided, it is necessary to increase the application of foliar fertilizer in a timely manner after the disaster, and appropriately increase the frequency of intertillage to raise the soil temperature, and promote the recovery and growth of cotton plants [17].

3.2.4.2 Rain disaster during the seedling stage

For unplanted drip-irrigated cotton fields, the “planting row delayed soil covering” technology is adopted, which means not covering the planting rows with soil during sowing to avoid soil compaction after rain. When the cotton plants reach the two-leaf and one-apical bud to three-leaf and one-apical bud stage, specialized soil covering machinery will be used for precise soil covering of the planting rows (**Figure 1**) [31].

3.2.4.3 Drought disaster

In response to the phenomenon of insufficient soil moisture or drought in cotton fields caused by incomplete irrigation before sowing when the suitable sowing period has arrived, a water supply strategy of sowing first and then using drip irrigation within 48 hours can be adopted.

3.2.4.4 Wind-sand hazard

Engineering protection: the outside of the cotton field adopts a grid (main forest belt combined with secondary forest belt), multi-species (trees, shrubs, herbs) strategy to strengthen the construction of windbreaks forest belts; the inside of the cotton field set up windproof soil pressure belts according to the change of the main wind direction, ensuring that they are perpendicular to the prevailing wind direction.

Agronomic measures: in response to the areas with high wind disaster-prone, the wheat-cotton simultaneous sowing technology can be promoted, utilizing the dwarf population formed by the rapid growth of wheat in the early stage to significantly reduce the damage of wind and sand to cotton seedlings (**Figure 2**).

3.2.4.5 Hail disaster

Relying on radar monitoring and intelligent warning platforms, artificial hail elimination operations should be carried out in a timely manner at the initial formation stage of hail cloud clusters.

In case of disaster, timely measures should be taken, including replanting or switching to other crops.

Figure 1.
The scene of the specialized soil covering machinery for "planting row delayed soil covering" operation.

Figure 2.
Wheat-cotton simultaneous sowing technology.

3.2.4.6 Early-frost during the late stage

Priority is given to cotton varieties with early maturing, and simultaneously implement cultivation measures to promote early maturity (the specific details are shown in Section 3.1).

3.2.4.7 Salt alkali disaster

Water conservancy salt discharge: carry out flood irrigation for salt washing before sowing and improve the field drainage system.

Soil improvement: carry out ultra-deep plowing (depth $\geq$ 40 cm) or ultra-deep tilling (depth $\geq$ 45 cm) once every 3 to 5 years. In areas where conditions permit, sand compaction measures can also be used to improve soil aggregate structure.

Biological improvement: Salt-tolerant green manure (such as sunflower) crop rotation; apply salt alkali-resistant microbial agents with drip irrigation water.

Chemical control: apply soil conditioners (humic acid-based or silicon-calcium-potassium-magnesium) with drip irrigation water.

Seed source prevention and control: strengthening the cultivation and promotion of cotton varieties with salt alkali-resistant property.

For the above-mentioned disasters: if the disaster occurs early, other extra-early-maturing cotton varieties can be planted according to soil moisture and nutrient conditions; if the disaster occurs late, alternative crops can be selected in a timely manner according to light and heat resources.

3.3 Mulched drip irrigation technology

The mulched drip irrigation technology is a deep combination of plastic film and mulching and drip irrigation systems. Its technical core lies in the coordinated layout of surface film mulching and drip irrigation tapes under film, achieving precise supply and efficient utilization of water and fertilizers; the technical advantage lies in the multiple effects of surface film covering, including increasing temperature and entropy preservation, suppressing salt and weed stress, promoting soil microbial activity. Practice has shown that the cotton fields using mulched drip irrigation technology achieve a 60% higher seedling survival rate compared with bare ground, while also accelerating the growth process and enabling the cotton flowering and boll-forming stages to highly coincide with the region's abundant light, heat, and water resources.

In practical production applications, the promoted film widths include 125, 145, 205, and 440 cm, which can flexibly adapt to different planting modes such as one film with four rows, one film with six rows, and even one film with twelve rows, among which the 205 cm film width is the most widely used. In addition, in order to ensure the effective recycling of residual film and reduce plastic film residue pollution, the film thickness of 0.01 mm or even 0.015 mm is commonly used in production (**Figure 3**).

At present, the promotion and application area of mulched drip irrigation technology combined with plastic film covering technology has reached 1.86 million hectares. The mulched drip irrigation system starts with a water source project, regulates water and fertilizer through the headworks hub, transports them through the water distribution pipeline network, and finally infiltrates the root zone of cotton evenly, regularly, and quantitatively with water and fertilizers through drip emitters. The advantage of this technology lies in that it can maintain the soil in the local range of cotton roots in a loose state and with the optimal water and fertilizer supply, effectively reduce interplant evaporation and deep leakage, and inhibit weed growth, fully realizing the comprehensive benefits of water conservation, fertilizer conservation, and yield increase.

In practical production applications, Xinjiang cotton fields usually drip 8–12 times during the whole growth period, and sandy soil cotton fields can add 1–2 more times. The first irrigation is generally carried out in early June, and for some cotton fields with heavy sand content, it can be advanced to late May or even earlier; the irrigation termination period is determined based on soil moisture, climatic conditions, and cotton growth status, generally at the end of August to stop irrigation, while sandy soil fields and prematurely aging cotton fields can postpone it to around September 10th. Each time drip quota is controlled at 300–525 m^3/ha, and total drip quota during the whole growth period is maintained at 2700–4200 m^3/ha. Topdressing generally

Figure 3.
Application of mulched drip irrigation technology in Xinjiang cotton fields.

follows the principle of "one irrigation with one fertilization", that is, when irrigating each time, N, P_2O_5, and K_2O fertilizers are applied synchronously by dripping. The specific details can be referred to **Table 3** [18, 27].

Due to the mulched drip irrigation technology in Xinjiang being an agricultural irrigation system with extremely distinctive regional characteristics, it has remarkable differences from conventional drip irrigation systems at home and abroad. Its main technological innovation features are reflected in the following aspects:

1. Technological integration innovation: combining film covering with drip irrigation system, relying on precision planter to complete the integrated operation of sowing, film laying, film pressing, and drip irrigation tape laying in one pass. For example, on a plastic film with a width of 205 cm, a drip irrigation tape layout of "two rows with one tape" is adopted, that is, planting six rows of cotton on the film, the film under the layout of three drip irrigation belts, which can be laid in the middle of the narrow rows (**Figure 4**) or on the side of the narrow rows (**Figure 5**).

2. Drip irrigation system design: Drip irrigation tapes are disposable consumables without pressure compensation. Under standard pressure (0.1 MPa), the flow rate of inlaid patch-type drip emitters is 1.8–3.2 L/h, and the flow rate of the single-wing labyrinth-type drip emitters is 2.0–3.4 L/h. The system operates at low pressure (0.03–0.07 MPa), and the filtration system uses a limited combination of filtration requirements that is simple and inexpensive but meets production requirements. Specifically, the surface water is filtered using a "sedimentation tank + non-pressure self-cleaning mesh filter + mesh/stacked filter" filtration system; the groundwater is filtered using a "centrifugal mesh/stacked filter" filtration system, and combined with the irrigation mode of large branch pipe rotational irrigation. This drip irrigation system is simple, durable, and highly adaptable. It not only ensures a high yield of cotton, but also reduces the cost from 5500 $/ha for conventional drip irrigation systems to 1300 $/ha, significantly reducing equipment investment costs, which has the dual advantages of "reducing costs and increasing yield" [11, 27].

Growth stage	Drip irrigation frequency	Drip irrigation date (month-day)	Drip quota m^3/ha	N (kg/ha)	P_2O_5 (kg/ha)	K_2O (kg/ha)
Bud stage	First time	6–6 ~ 6–19	300.0 ~ 525.0	18.0 ~ 22.5	7.5 ~ 15.0	—
	Second time	6–20 ~ 6–28	300.0 ~ 450.0	18.0 ~ 22.5	7.5 ~ 15.0	—
Flowering and boll stage	Third time	6–29 ~ 7–6	300.0 ~ 525.0	12.0 ~ 20.0	—	—
	Fourth time	7–7 ~ 7–13	300.0 ~ 525.0	12.0 ~ 20.0	—	—
	Fifth time	7–14 ~ 7–21	300.0 ~ 450.0	12.0 ~ 20.0	—	—
	Sixth time	7–22 ~ 7–30	300.0 ~ 450.0	12.0 ~ 20.0	—	—
	Seventh time	8–1 ~ 8–9	300.0 ~ 450.0	12.0 ~ 20.0	7.5 ~ 18.0	7.5 ~ 18.0
	Eighth time	8–10 ~ 8–17	300.0 ~ 450.0	12.0 ~ 20.0	7.5 ~ 18.0	7.5 ~ 15.0
Boll opening stage	Ninth time	8–18 ~ 8–29	300.0 ~ 375.0	12.0 ~ 21.0	—	—

Table 3.
Typical water and fertilizer management plan for drip irrigation cotton fields in Xinjiang.

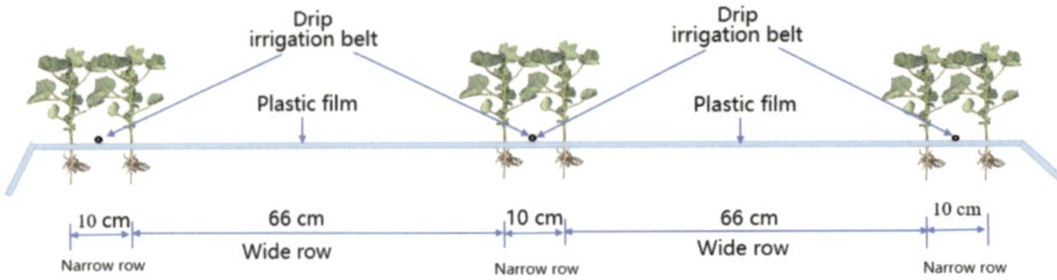

Figure 4.
Layout of cotton planting rows and drip irrigation belts (in the middle of narrow rows) on 205 cm plastic mulch.

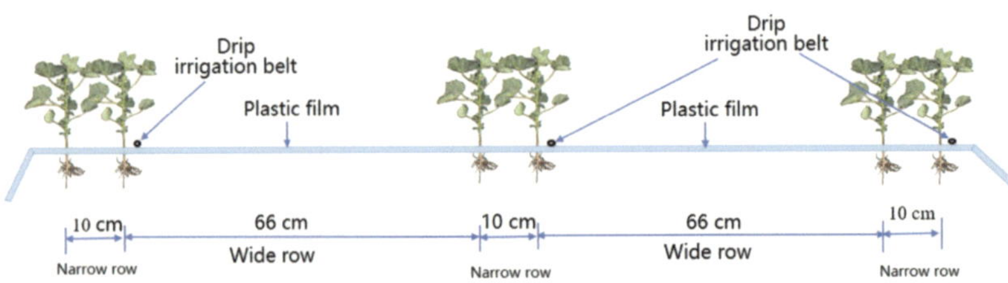

Figure 5.
Layout of cotton planting rows and drip irrigation belts (on the middle of narrow rows) on 205 cm plastic mulch.

The mulched drip irrigation technology in Xinjiang also shows significant advantages in water regulation. For example, Xinjiang has long faced the problem of seasonal water shortages in spring, resulting in many cotton fields not being irrigated, which makes soil moisture unable to meet the conventional requirements before sowing. After years of technological research and development, for cotton fields that have not undergone winter irrigation or spring irrigation before sowing, a strategy has been proposed to complete key operations such as tillage, soil preparation, sowing, and drip irrigation within 3 to 6 days before the arrival of suitable sowing period. Among them, the use of a split-flow type soil preparation machine, instead of a conventionally combined soil preparation machine in the soil preparation process, can effectively prevent the occurrence of issues, such as poor field flatness, seeding misplacement, and uneven sowing depth due to low flatness and loose soil after cotton field preparation; completely avoid the emergence of the problems, such as tractor sinking during sowing, uneven seedling, and poor seedling retention. For cotton fields that have not undergone winter irrigation or spring irrigation before sowing, it is clearly specified to carry out 1 to 2 times drip irrigation operations with a drip application rate of 180–450 m^3/ha after sowing. This strategy involves adjusting the large-quantity irrigation with a drip application rate of 1950–2475 m^3/ha from mid-February to March 15 before sowing to 1–2 times small-quantity irrigation with a drip application rate of 180–450 m^3/hm^2 around April 10 after sowing. It is specified that water should be dripped after sowing, with a special requirement that waterlogging on the surface is strictly prohibited. The drip intervals should be controlled at 5–7 days, and advocate the application of 15–30 kg/ha of humic acid fertilizer or the appropriate amount of biological fungicide along with the drip. For cotton fields with heavy salt alkali, soil conditioner should also be applied along with the drip. This strategy not only takes advantage of the time allocation of water resources but also makes better use of the favorable conditions of rapid temperature rise in spring, significantly increases river water resources, ensures the soil moisture required for cotton seedling growth, and greatly saves water resources. Currently, this technology's promotion and application area has exceeded 1 million hectares [32].

In addition, the mulched drip irrigation technology in Xinjiang can also effectively prevent and control the rain disaster during the seedling stage, that is, the "planting row delayed soil covering" technology is adopted, as detailed in the text given in Section 3.2.4.2 [31].

3.4 Population regulation technology

In the high-density and dwarfing planting mode of cotton fields in Xinjiang, improper population regulation and management can easily lead to field closure due to excessive cotton vegetative growth, or reduce yield due to inadequate construction of high-yield plant type. Therefore, after cotton emergence, on the basis of scientific implementation of water and fertilizer management, it is also necessary to fully utilize the chemical regulation effect of mepiquat chloride, that is, adhere to the strategy of "taking water and fertilizer regulation as the main approach and chemical regulation as the supplementary", with water and fertilizer regulation strategy details described in the text given in Section 3.3 mulched drip irrigation technology. Chemical regulation with mepiquat chloride is generally carried out 4–7 times during the whole growth period, including 1–3 times during the seedling and bud stage, and 2–5 times during the flowering and boll stage (including chemical topping). Among them, the dosage of mepiquat chloride during the seedling stage is about 15.0 g/ha per time; the chemical regulation of bud stage is generally carried out before the first drip irrigation (at the full bud stage or 5–7 real leaf stage), with a dosage of 22.5–37.5 g/ha per time; around flowering stage, especially in vigorously growing cotton fields, chemical regulation is usually carried out about 1 week before flowering, with a dosage of 5.0–60.0 g/hm^2 per time. If chemical regulation with mepiquat chloride is not carried out before flowering, it is generally carried out 2–5 days after the flowering stage in combination with drip irrigation, with a dosage of 45.0–67.5 g/hm^2 per time. Notably, during the whole cotton growth period, if the early-stage regulation effect is insufficient, one additional chemical regulation can be flexibly added at any time, with the dosage increased by about 30% on the basis of the conventional dosage; if the cotton field shows weak growth, drought stress, or excessive regulation, the frequency and dosage of mepiquat chloride should be reduced in a timely manner, and even the spraying of mepiquat chloride should be stopped [9, 24]. Mepiquat chloride chemical topping technology has recently been widely applied in production. It is generally carried out on July 10th (full flowering stage) of the year, spraying 25% mepiquat chloride water emulsion at a dosage of 40–60 mL/ha, or 98% mepiquat chloride soluble powder at a dosage of 180–225 g/ha, and using liquid additive, which is conducive to the uniform adhesion and absorption of the agent, with a dosage of 150 mL/ha. To ensure the effect of mepiquat chloride chemical topping, 98% mepiquat chloride soluble powder is often sprayed 3–7 days before chemical topping at a dosage of 45–75 g/ha, and then 98% mepiquat chloride soluble powder is sprayed 5–10 days after chemical topping for severe chemical control at a dosage of 150–270 g/ha. In addition, for cotton fields with vigorous growth in early and middle of April, one additional mepiquat chloride chemical regulation should be applied 4–5 days before dripping, with a dosage of 150–240 g/hm^2 [9].

3.5 Management techniques to match cotton machine harvesting

The mechanized cotton harvesting technology system in Xinjiang includes key links such as variety screening, plant spacing configuration, defoliation and ripening, cotton field preparation before harvesting, and mechanical harvesting. Each link has strict technical standard requirements, and the main contents are as follows:

3.5.1 Variety screening

Priority should be given to early maturity (with a growth period of about 125 days in southern Xinjiang and about 120 days in northern Xinjiang), high yield, and stress resistance; the distance between the lower opening boll and the ground is ≥18 cm; the plant type is moderately loose (to prevent cotton leaves from hanging branches); the upward-pointing leaves and leaf area of medium size; and the varieties are sensitive to defoliants. To ensure the quality of machine-harvested cotton, the fiber genetic quality indicators for selected upland cotton varieties should be as follows: 2.5% span length ≥ 30 mm, strength ≥30.0 cN/tex, uniformity ≥84.5%; the fiber genetic quality indicators for selected island cotton varieties should be as follows: 2.5% span length ≥ 37 mm, strength ≥45.0 cN/tex, uniformity ≥86.5%.

Based on the above-mentioned requirements, the recommended main upland cotton varieties in southern Xinjiang are Xinluzhong 84, Tahe 2, J206-5, Yuanmian 8, Xinta Cotton 3, J8031, etc.; the recommended main island cotton varieties are Xinhai 62, Xin 78, Changfeng 10, etc.; the recommended main upland cotton varieties in northern Xinjiang are Xinluzao 76, Xinluzao 78, Xinluzao 80, Xinluzao 84, H33-1-4, Jinken 1402, Zhongmian 113, F015-5, etc.; Northern Xinjiang is not suitable for planting island cotton varieties.

3.5.2 Plant spacing configuration

The current planting mode of cotton fields in Xinjiang includes two types of plant spacing configurations: Wide-narrow row spacing and equal row spacing. The details are as follows:

1. Wide-narrow row mode takes 66 + 10 cm as the basic standard (narrow row width ≤ 12 cm, the sum of the width of wide and narrow rows is 76 cm), derived row configurations include multiple combinations such as 64 + 12 cm, 64.5 + 11.5 cm, 65 + 11 cm, and 72 + 4 cm. This mode adopts precision single seed sowing, with the theoretical density controlled at 240,00 to 277,500 plants/hm^2.

2. Equal row spacing mode features a fixed row spacing of 76 cm. This mode adopts precision single seed sowing, with a density controlled between 145,000 and 234,000 plants/hm^2 [18].

3.5.3 Defoliation and ripening

The spraying time for defoliation and ripening should be determined based on the maturity status of cotton and climatic conditions. The defoliation and ripening promotion spraying should meet the following conditions: the natural opening rate of cotton fields should reach 40 to 50%; the upper cotton bolls should develop for 40 to 50 days; no rainfall within 24 hours after spraying; the minimum temperature stable at 10–15°C for 3 to 5 days, and the daily average temperature stable at 18–21°C for 7 to 10 days. The spraying operations for defoliation and ripening promotion are usually carried out from September 15 to 25 in southern Xinjiang, while they are carried out from September 5 to 15 in northern Xinjiang.

Selection and dosage determination of defoliation and ripening chemicals. When spraying with a tractor, use 54% (540 g/L) diuron thidiazuron suspension agent at a dosage of 180–225 mL/hm^2, combined with 40% ethephon water agent at a dosage of 1050–1500 mL/hm^2; or use 50% thidiazuron suspension agent at a dosage of 600–675 mL/hm^2. Two applications are generally required when spraying with drones. For the first time, use 540 g/L diuron·thidiazuron suspension agent at a dosage of 180–195 ml/hm^2, combined with 40% ethephon water agent at a dosage of 1050–1500 mL/hm^2; or use 50% thidiazuron suspension agent at a dosage of 450–600 mL/hm^2, combined with 50% dichlorvos powder at a dosage of 60–75 g/hm^2, and 40% ethephon water agent at a dosage of 1050–1350 ml/hm^2. For the second time, use 540 g/L diuron·thidiazuron suspension agent at a dosage of 90–120 ml/hm^2, combined with 40% ethephon water agent at a dosage of 600–900 ml/hm^2; or use 50% thidiazuron suspension agent at a dosage of 300 ml/hm^2, combined with 40% ethephon water agent at a dosage of 1050–1500 ml/hm^2 [9, 33].

3.5.4 Preparation in the field before harvesting

In order to ensure smooth and high-quality mechanized harvesting operations, a comprehensive inspection and preparation of the cotton field and its surrounding environment are required [33]. The specific operations are as follows:

1. Infrastructure inspection: ensure that the width of field roads and bridges is ≥4 m, the slope of roads and bridges is ≤30°, the height of overhead cables is ≥5 m, and remove obstacles or markings to ensure that cotton harvesting machinery can successfully avoid them.

2. Field cleaning: do not carry out plastic film and drip irrigation tape recycling operations before harvest, but timely recycle the horizontal branch pipes in the field, and bury the residual film, drip irrigation head or tail exposed outside the cotton field surface with soil, with the burial depth controlled at 6.5–8.5 cm, while removing weeds and floating objects (fragments of plastic film and waste drip irrigation tapes) in advance. After harvesting, extract the drip irrigation tapes and recycle the residual films.

Field entrance handling: manually harvest the boll opening cotton in areas about 15 meters wide at both ends of the cotton field entrance and difficult for machinery to harvest, in advance, to reduce the loss rate during mechanical harvesting [33].

3.5.5 Mechanical harvesting operation standards

1. Priority is given to six row self-propelled baler cotton harvests such as Boshiran, World Agricultural Machinery, Shandong Swan, and China Railway Construction Heavy Industry, as well as John Deere's CP770, CP690 cotton harvesters, and so on.

2. Mechanical harvesting cotton fields must meet the following requirements: the defoliation rate ≥ 90%, the boll opening rate ≥ 95%; strictly prohibited from

carrying out mechanical harvesting at night and at a time when the dew has not yet dried; harvesting time is recommended to be controlled between 12:00 and 21:00 daily. For some cotton areas in southern Xinjiang with relatively dry air, the harvesting time can be controlled in the day 11:30–24:00 daily.

3. During the mechanical harvesting process, the operator must strictly follow the operation procedures to reduce the seed cotton bumping down and avoid leakage and repeated harvesting. Mechanical harvesting standards are as follows: the net harvesting rate is stable at 93–95%, the total loss rate is ≤5%, the impurity rate is ≤12%, the moisture regain rate of seed cotton is ≤12%, and the vehicle speed is 3.5–4.0 km/h [12, 15].

3.5.6 Supporting management standards

1. Personnel configuration: operators of cotton harvesters and cotton transport vehicles must hold professional operating qualification certificates; reasonably configure the number of cotton transport vehicles based on factors such as the actual cotton field yield, transportation distance, and the work efficiency of cotton harvesters, four transport vehicles are generally allocated per cotton harvester.

2. Fire safety: each cotton picker must be equipped with one full-time firefighter, and the harvesting site must be equipped with no less than six fire extinguishers (8 kg ammonium phosphate salt type), one water tanker with a capacity of ≥4 m^3, and its supporting fire water pump; open flames are strictly prohibited within 100 meters of the harvesting operation area.

4. Conclusions

Since the twentieth century, the cotton planting area in the oases of Xinjiang has continued to expand. With the continuous improvement and promotion of planting technology, the unit yield and total production of cotton have continuously set new records. Xinjiang has become one of the most important cotton production areas in China and even the world.

At present, the level of cotton planting technology in Xinjiang has reached the forefront of the world, with some technologies reaching the international leading level. Specifically, cotton production in Xinjiang takes the "early, short, density" planting mode as a technological breakthrough, integrating key technologies such as comprehensive prevention and control of pests, diseases, weeds, and natural disasters, water-saving through mulched drip irrigation, and standardized management of machine-harvested cotton. These technologies not only lay a solid technical foundation for stable and high cotton yields but also fully demonstrate the concept of green development and the effectiveness of transforming scientific and technological achievements into practical productivity. Among them, the unique design of mulched drip irrigation technology is particularly suitable for oasis cotton areas with similar natural ecological conditions as Xinjiang. The enormous success of cotton production in Xinjiang's oases is essentially the result of high-efficiency adaptation between technological innovation and natural resources. Due to continuous technological innovations and breakthroughs, on the one hand, it has effectively addressed the challenges

DOI: http://dx.doi.org/10.5772/intechopen.1011390

of natural conditions in oasis areas such as high temperature, low temperature, drought, sandstorms, salt alkali, and poor water, heat, and soil resource; on the other hand, it has focused on fully exerting the advantages of Xinjiang's resource endowment, achieving the coordinated development of high and stable cotton yield, high quality, simple, and efficient. This practice has not only become a model for modern cotton cultivation in global oases, providing a "China Xinjiang cotton cultivation application plan" for the sustainable development of global cotton, but also making important contributions to the development of cotton cultivation science.

Author details

Liwen Tian[1†], Feng Shi[1†], Jianping Cui[1] and Honghai Luo[2*]

1 Cotton Research Institute of Xinjiang Uygur Autonomous Region Academy of Agricultural Sciences, Urumqi, Xinjiang, China

2 College of Agriculture, Shihezi University, Shihezi, Xinjiang, China

*Address all correspondence to: luohonghai79@163.com

† These authors contributed equally to this work.

References

[1] Guan QL, Liu WZ, Yu GX. Balance analysis of agricultural water and land resource in Xinjiang. Territory & Natural Resources Study. 2015;**2**:33-36. DOI: 10.3969/j.issn.1003-7853.2015.02.012. (in Chinese)

[2] Mao SC, Ma XF, Tian LW, et al. Research on the Sustainable Development of Cotton in Xinjiang Oasis. Shanghai: Shanghai Scientific and Technical Press; 2022. (in Chinese)

[3] Hu W, Yao L. The trend analysis on air temperature and evaporation in Kuche county during recent 40 years. Research of Soil and Water Conservation. 2008;**3**:93-95. DOI: CNKI:SUN:STBY.0.2008-03-023. (in Chinese)

[4] Zhu M, Liang Z, Xu WL, Zhou B, Ding F. Characteristics of spatial distribution in soil nutrition of cotton meld in Xinjiang oasis. Xinjiang Agricultural Sciences. 2009;**5**:1076-1081. DOI: CNKI:SUN:XJNX.0.2009-05-038. (in Chinese)

[5] Han CL. Temporal and spatial variation of soil nutrients of long-term monocultural cotton field and soil sustainable utilization in Xinjiang [thesis]. Shihezi: Shihezi University; 2010. (in Chinese)

[6] Tang MY, Lai B, Zhao Z. Investigation and evaluation of current situation of cotton fertilization in Xinjiang. Xinjiang Agricultural Science and Technology. 2016;**2**:48-50. DOI: 10.3969/j.issn.1007-3574.2016.02.027. (in Chinese)

[7] Cotton Research Institute of the Chinese Academy of Agricultural Sciences. Cultivation of Cotton in China. Shanghai: Shanghai Scientific and Technical Press; 2019. (in Chinese)

[8] Wan SM, Dong HL, Xu WX, Chen GD. Optimization of cotton field planting system in Xinjiang and creation of light and simplified techniques. China High and New Technology. 2022;**9**:146-148. DOI: 10.3969/j.issn.2096-4137.2022.09.065. (in Chinese)

[9] Tian LW, Aierxi A, Dong HZ, Kong J, Song MZ, Dong HL, et al. A High-Yield, High-Efficiency and Light-Simplified Cultivation Method for Cotton in Xinjiang Cotton Area. China: CN108293817A (P); 2018. (in Chinese)

[10] Nie JJ, Dai JL, Du MW, Zhang YJ, Tian XL, Li ZH, et al. New development of modern cotton farming theory and technology in China - concentrated maturation cultivation of cotton. Scientia Agricultura Sinica. 2021;**20**:4286-4298. DOI: 10.3864/j.issn.0578-1752.2021.20.004. (in Chinese)

[11] Cao PF, Liu ZX, Chen M, Song CJ. Current status, problems and recommendations of the application of drip irrigation technology in the Xinjiang production and construction corps. Water Resources Development Research. 2022;**5**:114-119. DOI: 10.13928/j.cnki.wrdr.2022.05.023. (in Chinese)

[12] China Daily. Science and Technology Have Created Miracles in Xinjiang's Cotton Production [Internet]. China Daily; 2022. Available from: https://cn.chinadaily.com.cn/a/202206/08/WS62a08122a3101c3ee7ad98e0.html [Accessed: June 8, 2022]. (in Chinese)

[13] Kong LY, Yu X. Research on the collaborative development of agricultural green governance and water-saving

irrigation in Xinjiang. Journal of Shihezi University (Philosophy and Social Sciences). 2022;**4**:32-39. DOI: 10.3969/j.issn.1671-0304.2022.04.006. (in Chinese)

[14] Li P, Cap H, Zhan W, Cao YB, Fan YS, Ma CY. Application mechanism of drip irrigation water saving technology and hot spot research Progress. Water Saving Irrigation. 2024;**2**(62-68):75. DOI: 10.12396/jsgg.2023369. (in Chinese)

[15] China Agricultural Mechanization Information Network [Market Observation]. The Domestic Cotton Picker Industry Enters a New Stage of High-Quality Development with Product Upgrading [Internet]. China Agricultural Mechanization Information Network; 2025. Available from: http://www.amic.agri.cn/secondLevelPage/info/314/203097.html [Accessed: April 23, 2025]. (in Chinese)

[16] Lu XR, Jia XY, Niu JH. The present situation and prospects of cotton industry development in China. Scientia Agricultura Sinica. 2018;**1**:26-36. DOI: 10.3864/j.issn.0578-1752.2018.01.003. (in Chinese)

[17] Lou SW, Tian LW, Luo HH, Du MW, Lin T, Yang T, et al. Analysis on key production techniques of cotton with good quality and high yield in Xinjiang. Scientia Agricultura Sinica. 2023;**14**:2673-2685. DOI: 10.3864/j.issn.0578-1752.2023.14.004. (in Chinese)

[18] Tian LW, Lou SW, Zhang PZ, Du MW, Luo HH, Li J, et al. Analysis on the problems and approaches for improving the unit yield of cotton in Xinjiang under the green and efficient production model. Scientia Agricultura Sinica. 2025;**6**:1102-1115. DOI: 10.3864/j.issn.0578-1752.2025.06.005. (in Chinese)

[19] National Bureau of Statistics. Announcement of the National Bureau of Statistics on Cotton Production in 2023 [Internet]. National Bureau of Statistics; 2023. Available from: https://www.stats.gov.cn/xxgk/sjfb/zxfb2020/202312/t20231225_1945749.html [Accessed: December 25, 2023]. (in Chinese)

[20] National Bureau of Statistics. Announcement of the National Bureau of Statistics on Cotton Production in 2024 [Internet]. National Bureau of Statistics; 2024. Available from: https://www.stats.gov.cn/sj/zxfb/202412/t20241225_1957879.html [Accessed: December 25, 2024]. (in Chinese)

[21] Mao SC, Wang GQ, Wang L, et al. Technology of High-Quality Cotton Production Chain in Xinjiang Oasis. Urumqi: Xinjiang Science and Technology Press; 2024. (in Chinese)

[22] Mao SC, Yu XX, Ge Q, Wang XJ, Zheng Y, Mai WX, et al. Research on the current situation of high-quality cotton production in Xinjiang oasis and countermeasures for accelerating the construction of high-quality cotton planting belts (I). China Cotton. 2024;**51**(5):1-8. DOI: 10.11963/cc20240010. (in Chinese)

[23] Tian LW, Aierxi A, Dong HZ, Kong J, Song MZ, Dong HL, et al. A Method for Constructing a Suitable Machine - Harvested Population in Drip - Irrigated Cotton Fields in the Tarim Basin of Southern Xinjiang. China: CN106856969A (P); 2017. (in Chinese)

[24] Tian LW, Dong HZ, Lou SW, Ning XM, Wang CW, Jiang CJ, et al. Analysis of the key cultivation techniques of "short, dense and early" under the light-simplified and efficient strategy for cotton in Xinjiang. In: Proceedings of the 2017 Annual Conference and 9th Membership Representative Assembly of

the Cotton Branch, Chinese Association of Agricultural Science Societies; 8-9 August 2017; Zheng Zhou, China. Cotton Branch, Chinese Association of Agricultural Science Societies; 2017. pp. 111-117. (in Chinese)

[25] Lou SW, Dong HZ, Tian XL, Tian LW. The "short, dense and early" cultivation of cotton in Xinjiang: History, current situation and prospect. Scientia Agricultura Sinica. 2021;**4**:720-732. DOI: 10.3864/j.issn.0578-1752.2021.04.005. (in Chinese)

[26] Tian LW, Cui JP, Guo RS, Xu HJ, Lin T, Liu SJ, et al. Cotton Field Seedling Protection Method with Precision Sowing. China: CN103404357A (P); 2013. (in Chinese)

[27] Luo HH, Tian LW, Dong HZ, Zhang N, Liu LT, Feng GP, et al. A Low-Cost and High-Efficiency Drip Irrigation System for Cotton Fields and its Usage Method. China: CN115474534A (P); 2022. (in Chinese)

[28] Tian LW, Ning XM, Wang GF, Wei X, Xu HJ, Cui JP, et al. Seed Selection Equipment. China: CN208554990U (P); 2019. (in Chinese)

[29] Dong HZ, Yang GZ, Tian LW, et al. Light and Simplified Cultivation of Cotton. Beijing: Science Press; 2016. pp. 291-303. (in Chinese)

[30] Xinjiang Uygur Autonomous Region Market Supervision and Administration Bureau. Technical Regulations for the Integrated Control of Major Cotton Diseases and Pests. China: DB65/T2271-2020(S); 2020. (in Chinese)

[31] Tian LW, Aierxi A, Lei B, Kong J, Chen Z, Zhao Q, et al. A Method for Protecting Cotton Seedlings in Fields without Pre-Sowing Irrigation under the Cultivation Environment of Precision Sowing with Drip Irrigation under Plastic Film. China: CN108157105A (P); 2018. (in Chinese)

[32] Tian LW, Xu HJ, Kong J, Lin T, Ning XM. Optimization and analysis of countermeasures for sustainable development of cotton in Xinjiang. China Fiber Inspection. 2018;**9**:110-113. DOI: 10.14162/j.cnki.11-4772/t.2018.09.031. (in Chinese)

[33] Tian LW, Liu LT, Nie XH, Song MZ, Zhang YB, Luo Z, et al. A Production Method for High-Quality Machine-Harvested Cotton in Xinjiang. China: CN111713365A (P); 2020. (in Chinese)

Chapter 3

Sustainable Smart Irrigation for Efficient Cotton Production

Marimuthu Subramani, Sureshkumar Rajalingam, Jeyajothi Raman and Sivakumar Kaliyannan

Abstract

Sustainable irrigation is essential for improving water use efficiency in cotton farming while ensuring long-term environmental and economic viability. This chapter explores the integration of smart irrigation technologies, such as IoT-based systems, AI-driven water management, and precision irrigation techniques, to optimize water usage in cotton production. Key strategies, including drip and subsurface irrigation, soil moisture monitoring, and remote sensing applications, are discussed to highlight their role in reducing water waste and enhancing crop yields. Additionally, water conservation methods such as rainwater harvesting, deficit irrigation, and the use of recycled water are examined as sustainable alternatives to conventional practices. The chapter also addresses the economic and environmental benefits of adopting smart irrigation, including lower water consumption, improved soil health, and increased profitability for farmers. By leveraging data-driven irrigation systems and climate-adaptive techniques, sustainable smart irrigation presents a transformative approach to efficient cotton farming, contributing to global efforts in water conservation and sustainable agriculture.

Keywords: smart irrigation, water use efficiency, drip irrigation, IoT-based irrigation, cotton yield

1. Introduction

Sustainable irrigation is critical for increasing water efficiency in cotton agriculture while also promoting long-term environmental sustainability and economic viability. With increasing water shortages and rising cotton demand, it is critical to implement innovative irrigation systems that maximize resource utilization. Traditional irrigation methods, such as flood irrigation, are no longer adequate to address the present issues of climate change, population development, and water shortage [1]. As a result, the agriculture sector is undergoing major transition, with an increasing emphasis on integrating technology to optimize input utilization and reduce environmental effect. Smart irrigation, a concept that combines classic agronomic principles with cutting-edge digital tools, has emerged as a game-changing strategy for effective cotton production [2]. It entails the use of automated systems, sensor-based monitoring, real-time data analysis, and remote irrigation control access, all of which help to improve informed decision-making and adaptive water management. These technologies allow farmers to conserve water, save energy, and increase crop output by matching irrigation schedules to real crop water requirements.

Cotton plants (*Gossypium spp.*) have a high water requirement throughout their life cycle, from germination to boll formation. Inefficient water application can cause water stress, poorer yields, worse fiber quality, and higher susceptibility to pests and diseases. To overcome these issues, sustainable irrigation systems must provide not just effective water delivery but also long-term soil health, ecosystem balance, and economic rewards to farmers [3]. Sustainable irrigation also includes water conservation measures such as rainwater collecting, shortfall irrigation, mulching, and treated wastewater reuse. These techniques, together with smart technologies, form a comprehensive foundation for water-efficient agriculture. The economic ramifications are as enormous, with farmers benefiting from lower input costs, higher yields, and better resource allocation [4].

This chapter investigates the use of smart irrigation technologies in cotton farming, including Internet of Things (IoT)-based systems, artificial intelligence (AI)-driven water management, and precision irrigation techniques, to optimize water utilization. These advanced tools allow for site-specific irrigation scheduling using real-time data from soil moisture sensors, weather stations, and crop growth models [5]. As a result, the irrigation process becomes more dynamic and responsive to changing environmental circumstances, resulting in better crop production while preserving natural resources.

Furthermore, the chapter looks at essential irrigation systems including drip and subsurface drip irrigation (SDI), which are known for their excellent water efficiency. These technologies supply water directly to the root zone, reducing evaporation and increasing nutrient uptake. The use of soil moisture monitoring technologies like tensiometers and capacitance probes improves irrigation precision by providing reliable data on soil water availability. Another important topic highlighted is the use of remote sensing and geographic information systems (GIS) in irrigation management [6]. Satellite imagery and drone-based mapping technologies assist farmers in determining crop health, soil moisture distribution, and water stress levels across wide areas. When combined with decision support systems (DSS), these tools can direct focused irrigation actions while reducing waste [7].

Finally, sustainable smart irrigation provides a forward-thinking answer to the water difficulties of cotton agriculture [8]. Farmers may increase water productivity, ensure long-term soil fertility, and generate higher economic returns by implementing data-driven, climate-resilient irrigation strategies. The integration of digital agriculture with sustainable water management not only strengthens farm resilience but also contributes to regional and global efforts to conserve water resources and ensure food and fiber security [9, 10]. The following sections of this chapter will delve into the specific components of smart irrigation systems, such as technological innovations, practical implementation strategies, and real-world case studies that show their effectiveness in increasing cotton production under a variety of agro-climatic conditions.

2. Importance of water use efficiency in cotton

Water Use Efficiency (WUE) is a vital parameter in cotton farming, particularly in water-scarce and drought-prone regions. It measures the yield or biomass produced per unit of water consumed. Enhancing WUE ensures better productivity and sustainability by maximizing the utility of available water resources [11]. Traditional irrigation methods such as surface or flood irrigation often result in high water losses via evaporation, runoff, and deep percolation, reducing irrigation efficiency and causing environmental issues such as salinity, nutrient leaching, and soil degradation [12, 13].

Modern cotton cultivation requires precise irrigation management that aligns water supply with the plant's growth stages—germination, vegetative phase, flowering, boll development, and boll opening. Technologies such as soil moisture sensors, crop models, and weather-based irrigation scheduling help in optimizing water input. Improved WUE supports better nutrient uptake, robust root systems, and greater plant resilience, enhancing both yield and lint quality while reducing chemical input dependency [9]. Economically, higher WUE reduces irrigation costs, mitigates water scarcity risks, and improves income stability for farmers. Government support through subsidies, training, and incentives further drives the adoption of water-saving technologies in cotton farming [14, 15].

2.1 Challenges of water management in cotton ecosystems

Cotton's high water demand, coupled with variable rainfall and rising temperatures, creates pressure on water availability. Conventional surface irrigation has low application efficiency due to evaporation, percolation, and runoff. Uneven water distribution also leads to plant stress and variable growth within the same field [16]. Infrastructure issues like unlined canals and inefficient pumps, along with over-extraction of groundwater, compound the problem—leading to declining water tables and increased costs.

Socioeconomic barriers, such as limited awareness, financial constraints, and weak institutional support, also hinder efficient water use. Many farmers hesitate to adopt new irrigation technologies due to perceived high costs or lack of technical know-how [11, 17]. Addressing these challenges requires an integrated approach involving infrastructure upgrades, farmer education, supportive policies, and stakeholder collaboration. Promoting advanced irrigation systems and data-driven tools is key to overcoming these barriers and ensuring sustainable cotton production [18, 19].

2.2 Critical growth stages and crop water requirements

Water needs vary across the cotton growth cycle. During germination and seedling stages, adequate soil moisture is essential for uniform emergence. In the vegetative phase, moderate water is needed to ensure healthy canopy development without waterlogging. The reproductive stages—squaring and flowering—are highly sensitive to stress, with water deficits leading to flower and boll drop [13]. The boll development phase demands the most water, critical for fiber elongation and yield formation. Toward the end of the season, limited irrigation during boll opening helps prevent excessive vegetative growth and facilitates timely harvest.

By aligning irrigation schedules with these critical stages, farmers can reduce wastage and improve yield outcomes. Tools like soil moisture probes, evapotranspiration models, and satellite-based monitoring support informed irrigation decisions, boosting WUE and crop health [20].

2.3 Role of technological interventions in enhancing WUE

Technologies such as soil moisture sensors (tensiometers, gypsum blocks), remote sensing, and weather-based irrigation tools allow precise monitoring of soil and crop water status. ET-based scheduling applies water only when the crop needs it [21]. Remote sensing and drones offer real-time visual maps of crop stress, which, when integrated with GIS and GPS-based systems, enable targeted irrigation through Variable Rate Irrigation (VRI).

IoT-enabled smart irrigation systems combine sensor, climate, and crop data, allowing remote monitoring and automated irrigation scheduling. Artificial Intelligence (AI) algorithms help forecast crop water needs and generate optimal watering plans. Mobile apps and decision support tools provide farmers with alerts, advisories, and analytics—making precision water management more accessible [22]. Together, these technologies not only enhance WUE but also improve overall farm efficiency, reduce energy use, and support environmentally sustainable cotton production.

3. Smart irrigation technologies for cotton

The integration of smart irrigation technologies in cotton farming represents a transformative step toward achieving sustainable water management and maximizing crop productivity. These technologies incorporate automation, real-time monitoring, precision control, and data-driven decision-making to optimize water use efficiency (WUE). In cotton cultivation, especially in arid and semi-arid regions, these systems help mitigate the challenges posed by water scarcity, climate variability, and inefficient traditional irrigation methods.

3.1 Drip and subsurface drip irrigation systems

Drip irrigation delivers water directly to the root zone of plants through a network of valves, pipes, tubing, and emitters. This method minimizes evaporation losses and ensures uniform water application. In subsurface drip irrigation (SDI), water is applied below the soil surface through buried drip lines, further reducing surface evaporation and weed growth [23].

Studies have shown that drip and SDI systems can improve WUE by 30–50% in cotton compared to traditional surface irrigation methods [2]. These systems also allow for fertigation, enabling the simultaneous application of nutrients, which enhances nutrient uptake and crop growth [24].

Despite their benefits, the initial cost of installation and maintenance can be a barrier, which necessitates institutional support through subsidies and training.

3.2 Soil moisture sensors and monitoring systems

Soil moisture sensors play a critical role in precision irrigation by measuring the amount of water available in the root zone. Common types include: Tensiometers, Capacitance sensors, and Time-domain reflectometry (TDR) sensors [23].

These sensors are often integrated with data loggers and communication modules, providing real-time soil moisture status via mobile or web platforms [25, 26]. This helps farmers decide when and how much to irrigate, avoiding over- or under-watering.

By delivering water precisely when and where it is needed, these technologies ensure that the crop maintains optimal physiological function throughout its lifecycle. Additionally, smart systems significantly contribute to the reduction of water wastage by minimizing losses due to over-irrigation, evaporation, and runoff. This targeted approach also aids in the prevention of nutrient leaching and waterlogging, common issues in conventional irrigation practices that degrade soil health and reduce crop productivity [23, 24]. Moreover, the ability to monitor and respond to field variability allows for tailored irrigation to specific zones, addressing microclimatic differences and soil heterogeneity within a field.

Advanced systems are capable of automating irrigation schedules based on preset moisture thresholds, further improving operational efficiency and crop performance [27].

3.3 IoT-based smart irrigation systems

The Internet of Things (IoT) plays a transformative role in modern cotton irrigation by integrating a network of sensors, actuators, controllers, and mobile devices to remotely monitor and manage irrigation systems [28]. These interconnected components continuously collect critical environmental and agronomic data such as soil moisture, air temperature, relative humidity, solar radiation, and wind speed, providing farmers with a comprehensive understanding of their field conditions. Through centralized IoT platforms, real-time data are displayed on user-friendly dashboards accessible via smartphones or computers, allowing for the remote activation of pumps, valve control, and the receipt of alerts regarding anomalies or malfunctions [15, 29].

Practical applications of IoT in cotton irrigation include remote-controlled valve operations driven by live sensor readings, automated alerts for equipment failures or unusual field conditions, and smart fertigation units that precisely deliver water and nutrients based on crop requirements [30]. These innovations dramatically enhance resource use efficiency, reduce human labor, and ensure optimal crop performance, particularly in large-scale farms or fields characterized by spatial variability [31]. By integrating IoT, cotton producers can shift from traditional irrigation schedules to responsive, data-driven strategies, promoting both sustainability and profitability in water-limited environments [1].

3.4 Artificial intelligence and decision support systems

Artificial Intelligence (AI) is revolutionizing irrigation management in cotton production by analyzing large volumes of data collected from field sensors, weather forecasts, and historical crop performance records. Machine learning models within AI systems can accurately predict evapotranspiration (ET) and estimate crop water requirements, thereby enabling precise irrigation scheduling. These models also help recommend optimal irrigation timing and volume to match the specific needs of the cotton crop at different growth stages. In addition, AI can detect anomalies such as irregular plant responses or potential pest and disease outbreaks that may be linked to irrigation inefficiencies [24, 27].

Decision Support Systems (DSS) equipped with AI tools offer user-friendly platforms that translate complex datasets into actionable insights for farmers. These systems support dynamic irrigation planning, allow for real-time risk assessments—such as forecasting drought conditions—and facilitate scenario analysis to evaluate different irrigation strategies under varying environmental and economic conditions [19]. For example, AI-based models can calculate the most effective irrigation frequency based on factors like soil type, cotton genotype, and local climate, ensuring that water is applied only when necessary and in the most effective manner. This level of precision promotes both resource efficiency and yield maximization, making AI a powerful tool in the drive toward sustainable smart irrigation in cotton farming [26, 29].

3.5 Remote sensing and GIS applications

Remote sensing technologies, including satellite imagery and Unmanned Aerial Vehicles (UAVs or drones), have emerged as powerful tools in smart irrigation for cotton production [32]. These technologies generate high-resolution, multispectral data that allow for real-time assessment of key crop and environmental parameters, such as crop water stress, canopy temperature, and vegetation indices like NDVI (Normalized Difference Vegetation Index) and NDWI (Normalized Difference Water Index). When coupled with Geographic Information Systems (GIS), remote sensing facilitates detailed spatial analysis and mapping, laying the foundation for Variable Rate Irrigation (VRI)—an advanced technique that delivers water differentially across a field based on soil type, crop health, and moisture status [29, 33].

In cotton farming, these technologies have specific applications including the early detection of drought stress, precision delineation of irrigation zones, and analysis of field variability for site-specific irrigation management [9]. By integrating data from satellites and drones into their irrigation strategies, farmers can achieve more efficient water use, enhanced yield uniformity, and reduced input costs while simultaneously improving the adaptability of cotton production systems to changing environmental conditions [34].

3.6 Mobile apps and farmer-friendly interfaces

Mobile-based applications and digital platforms are playing a pivotal role in democratizing smart irrigation technologies by making them more accessible to farmers, especially smallholders [35]. These user-friendly apps provide a range of functionalities such as field-level monitoring, localized weather forecasts, irrigation alerts, and interactive dashboards for data visualization. They often include step-by-step tutorials, FAQs, and advisory services to guide farmers in optimizing irrigation practices [36, 37].

Notable examples include mKrishi®, developed by Tata Consultancy Services, which offers customized advisories based on crop and soil data; IrrigatorPro, which supports precise irrigation scheduling for cotton based on soil and weather data; and Kisan Suvidha, launched by the Government of India, which provides information on weather, market prices, input dealers, and advisory services [31].

These platforms significantly enhance farmer decision-making by providing real-time, localized insights and actionable recommendations, thus bridging the technological divide and encouraging wider adoption of smart irrigation in cotton cultivation [9, 38].

3.7 Integration with renewable energy for irrigation

Smart irrigation systems can be effectively powered by solar energy, offering a sustainable solution for cotton farms located in off-grid or remote regions. The integration of solar-powered drip irrigation systems not only ensures reliable water delivery but also significantly reduces dependence on fossil fuels, thereby cutting energy costs and contributing to environmental sustainability [39, 40].

Among the key advantages of these systems are their uninterrupted operation during power outages, lower greenhouse gas emissions, and their role in promoting a sustainable energy-water-agriculture nexus. This synergy is particularly important for climate-resilient agriculture [27]. Countries like India have witnessed a growing adoption of solar-powered smart irrigation, driven by supportive government policies, subsidies, and renewable energy initiatives, which are enabling small and marginal farmers to embrace these technologies more widely [10, 33].

4. Case studies

4.1 India: Solar-powered smart irrigation in Maharashtra

In Maharashtra's drought-prone regions, particularly in Vidarbha and Marathwada, solar-powered drip irrigation systems have been successfully implemented for cotton farmers under the "National Solar Pumping Programme for Irrigation". These systems integrate IoT-enabled drip irrigation with solar energy, allowing farmers to monitor soil moisture and schedule irrigation remotely via mobile apps. A pilot project by Tata Power Solar in collaboration with IIT Bombay demonstrated a 30–40% improvement in water use efficiency and 20% increase in cotton yield compared to traditional practices [41].

Furthermore, farmers saved significantly on electricity costs and reported fewer instances of waterlogging and salinity buildup. The success of this model has prompted similar programs in Rajasthan and Gujarat under the KUSUM scheme promoting solar irrigation pumps.

4.2 Tamil Nadu: IoT-based irrigation advisory by TNAU

The Tamil Nadu Agricultural University (TNAU) developed an IoT-based irrigation advisory system for cotton and other field crops. By deploying soil moisture sensors, weather stations, and mobile alerts, the system provided customized irrigation recommendations based on evapotranspiration and soil water deficit levels. Field trials in Coimbatore and Erode districts showed that farmers were able to reduce irrigation frequency by 25% without yield penalties while also lowering input costs and enhancing sustainability [17].

4.3 United States: IrrigatorPro for precision cotton irrigation

In the Southeastern United States, particularly in Georgia and Mississippi, the USDA-ARS developed a smart irrigation app called IrrigatorPro, tailored specifically for cotton farmers. This system uses real-time field data and machine learning models to recommend optimal irrigation timing and volume. Studies reported up to 50% reduction in water use and a 10–15% improvement in lint yield when compared with conventional irrigation schedules [42].

This tool is freely accessible and has been adopted by large-scale commercial cotton growers to reduce resource wastage, improve yield uniformity, and meet environmental regulations.

4.4 Australia: Smart scheduling with remote sensing in New South Wales

In New South Wales, the Cotton Research and Development Corporation (CRDC) has partnered with AgTech companies to implement remote sensing and GIS-based Variable Rate Irrigation (VRI) systems in large cotton farms. These systems integrate drone and satellite imagery with real-time weather data to identify areas of high evapotranspiration and target irrigation accordingly. This zone-specific irrigation approach resulted in 15% water savings and increased fiber uniformity, which is critical for export-quality cotton [43].

5. Challenges and future perspectives for smart irrigation in cotton

5.1 Challenges in adopting smart irrigation for cotton

Despite the numerous benefits of smart irrigation technologies, their widespread adoption in cotton farming—especially in developing countries—faces several technical, economic, infrastructural, and institutional challenges [44].

5.1.1 High initial investment and operational costs

Smart irrigation systems such as drip irrigation, IoT sensors, weather stations, and AI-powered tools require significant upfront investments. Small and marginal cotton farmers often lack the capital to afford these technologies, even with subsidies [31]. In addition, recurring costs for maintenance, internet connectivity, and software subscriptions can further burden economically vulnerable farmers [45].

5.1.2 Limited awareness and technical knowledge

A major bottleneck is the lack of awareness and training among farmers regarding the installation, use, and maintenance of advanced irrigation tools. Many farmers are unfamiliar with interpreting sensor data, managing digital dashboards, or acting upon app-based recommendations. Without adequate capacity building, even subsidized technologies remain underutilized or misused [44].

5.1.3 Inadequate infrastructure and connectivity

Reliable power supply, internet connectivity, and mobile network coverage are essential for operating IoT- and AI-driven irrigation systems. However, in many cotton-growing regions—especially in Sub-Saharan Africa and parts of South Asia—basic infrastructure is still lacking. This restricts the real-time functioning and remote control of irrigation systems [46].

5.1.4 Data privacy and technology standardization

With the increasing use of sensors, cloud storage, and AI tools, data ownership and privacy concerns are emerging. Farmers may be reluctant to share their farm data without guarantees on confidentiality and fair use. Moreover, the lack of standardization across irrigation tools and platforms leads to compatibility issues and discourages widespread adoption [47].

5.1.5 Environmental and agronomic concerns

Improper implementation of smart irrigation technologies can lead to over-irrigation in certain zones, malfunctioning sensors, or imbalanced fertigation if the data inputs are inaccurate. Also, many tools are not calibrated for region-specific cotton varieties, soil types, or climatic zones, which can reduce effectiveness if not locally adapted [43, 48].

5.2 Future perspectives and opportunities

Despite these challenges, the future of smart irrigation in cotton production is promising. Innovations in digital agriculture, along with policy support and increased investment, are opening new opportunities for scalable, inclusive, and sustainable water management.

5.2.1 Integration of AI, IoT, and blockchain

The convergence of Artificial Intelligence (AI), Internet of Things (IoT), and blockchain technology will enhance transparency, precision, and security in irrigation management. AI-driven decision-making combined with real-time sensor data will provide customized irrigation schedules, while blockchain can ensure secure data sharing and traceability [44, 49].

5.2.2 Low-cost, open-source technologies

Startups and research institutions are increasingly focusing on developing affordable and open-source irrigation solutions tailored for smallholders. These include DIY soil moisture kits, SMS-based advisory platforms, and solar-integrated smart pumps, which reduce entry barriers and democratize access to precision irrigation [50].

5.2.3 Public-private partnerships (PPP) and policy support

Collaborations between government agencies, private companies, and agricultural universities can promote the mainstreaming of smart irrigation systems. Subsidies under programs like India's PM-KUSUM, USA's EQIP, or Australia's Smart Farms program are examples of how policy can scale up adoption [51].

5.2.4 Climate resilience and sustainability goals

With rising climate variability, smart irrigation will play a central role in climate-smart agriculture by conserving water, reducing emissions, and enhancing soil health [34, 52]. Technologies that integrate weather forecasting, carbon footprint monitoring, and adaptive irrigation scheduling will contribute to achieving SDGs, especially Goal 6 (clean water) and Goal 13 (climate action).

5.2.5 Farmer-centric innovation models

Future innovations must focus on user-friendly, language-localized, and context-specific tools designed with and for farmers. Participatory approaches, on-field demonstrations, and digital literacy programs can help build trust and confidence in smart irrigation [53].

6. Conclusion

Water is a critical input in cotton cultivation, especially in the face of increasing water scarcity, erratic rainfall patterns, and growing competition for freshwater resources. As a high-input and water-intensive crop, cotton presents both a challenge and an opportunity for sustainable water management through the adoption of smart irrigation technologies. The integration of tools such as drip systems, soil moisture sensors, IoT platforms, AI-driven decision support systems, GIS mapping, and mobile applications has revolutionized the way farmers can manage irrigation—allowing for precise, efficient, and adaptive practices tailored to the crop's needs and the field's variability. This chapter has emphasized the importance of Water Use Efficiency (WUE) in cotton and explored the challenges associated with traditional irrigation methods. Smart irrigation strategies not only reduce water wastage and improve yield but also contribute to enhanced crop health, soil conservation, and reduced greenhouse gas emissions. Through case studies from India and abroad, it is evident that smart irrigation is not merely a technological advancement but also a pathway to climate-resilient and economically viable cotton production. However, widespread adoption is still hindered by barriers such as high initial costs, limited farmer awareness, lack of infrastructure, and policy gaps. Moving forward, addressing these challenges through multi-stakeholder collaborations, increased public–private partnerships, capacity building, and region-specific research and innovation will be essential. Furthermore, ensuring inclusivity by designing affordable, scalable, and farmer-friendly solutions can help bridge the digital divide and make smart irrigation accessible to smallholders who form the backbone of cotton farming in many countries. In conclusion, smart irrigation stands at the intersection of technology, sustainability, and agricultural transformation. Its judicious implementation can significantly contribute to achieving higher productivity, resource conservation, and environmental resilience, making it an indispensable tool in the future of cotton cultivation.

Author details

Marimuthu Subramani[1*], Sureshkumar Rajalingam[2], Jeyajothi Raman[1] and Sivakumar Kaliyannan[1]

1 Department of Agronomy, SRM College of Agricultural Sciences - SRMIST, Chengalpattu, Tamil Nadu, India

2 Department of Agronomy, Amrita School of Agricultural Sciences, Amrita Vishwa Vidyapeetham, Coimbatore, Tamil Nadu, India

*Address all correspondence to: marimuts@srmist.edu.in

References

[1] McCarthy A, Foley J, Raedts P, Hills J. Field evaluation of automated site-specific irrigation for cotton and perennial ryegrass using soil-water sensors and model predictive control. Agricultural Water Management. 2023;**277**:1-17

[2] Vellidis G, Tucker M, Perry C, Reckford D, Butts C, Henry H, et al. Irrigation scheduling for cotton using soil moisture sensors: University of Georgia SmartIrrigation Cotton App and sensor array. In: Beltwide Cotton Conferences. Camilla, GA; 2016

[3] Ahmad HS, Imran M, Ahmad F, Rukh S, Ikram RM, Rafique HM, et al. Improving water use efficiency through reduced irrigation for sustainable cotton production. Sustainability. 2021;**13**(7):4044

[4] Wan L, Jian Y, Zhang M, Tong J, Rebi A, Zhou J. Improving water productivity using subsurface drip irrigation in Yunnan province. Agronomy. 2024;**14**(4):679

[5] Mohammad Kunt YE. Development of a smart autonomous irrigation system using IoT and AI. arXiv preprint; 2025

[6] Bin L, Shahzad M, Khan H, Bashir MM, Ullah A, Siddique M. Sustainable smart agriculture farming for cotton crop: A fuzzy logic rule based methodology. Sustainability. 2023;**15**(18):13874

[7] Ghaffar A, M H u R, Ali HR, Haider G, Ahmad S, Fahad S, et al. Modern concepts and techniques for better cotton production. In: Cotton production and uses: Agronomy, crop protection, and postharvest technologies. Singapore: Springer Singapore; 2020. pp. 589-628

[8] Ali A, Tajamul H, Azlan Z. Smart irrigation technologies and prospects for enhancing water use efficiency for sustainable agriculture. AgriEngineering. 2025;7(4):106

[9] Ahmed Z, Gui D, Murtaza G, Yunfei L, Ali S. An overview of smart irrigation management for improving water productivity under climate change in drylands. Agronomy. 2023;**13**(8):2113

[10] Wang L, He L, Sun W, Gao C, Han Z, Lin M. Precise irrigation of dryland cotton under canal irrigation system constraints based on the CERES-CROPGRO-cotton model. Agricultural Water Management. 2025;**317**:109624

[11] Mollaee M, Mobli A, Mutti NK, Manalil S, Chauhan BS. Challenges and opportunities in cotton production. Cotton production. 2019;**13**:371-390

[12] Datta A, Ullah H, Ferdous Z, Santiago-Arenas R, Attia A. Water management in cotton. Cotton production. 2019;**13**:47-59

[13] Ali MH. Crop water requirement and irrigation scheduling. In: Fundamentals of irrigation and on-farm water management. Vol. 1. New York, NY: Springer New York; 2010. pp. 399-452

[14] Luo Q, Bange M, Johnston D, Braunack M. Cotton crop water use and water use efficiency in a changing climate. Agriculture, Ecosystems & Environment. 2015;**202**:126-134

[15] Chen Y, Yu Z, Han Z, Sun W, He L. A decision-making system for cotton irrigation based on reinforcement learning strategy. Agronomy. 2023;**14**(1):11

[16] Zhang Z, Huang J, Yao Y, Peters G, Macdonald B, La Rosa AD, et al. Environmental impacts of cotton and opportunities for improvement. Nature Reviews Earth & Environment. 2023;**4**(10):703-715

[17] Tamil Nadu Agricultural University (TNAU). IoT-Based Irrigation Scheduling Systems. Coimbatore: Department of Agronomy; 2022

[18] Koudahe K, Sheshukov AY, Aguilar J, Djaman K. Irrigation-water management and productivity of cotton: A review. Sustainability. 2021;**13**(18):10070

[19] Uddin J, Smith RJ, Gillies MH, Moller P, Robson D. Smart automated furrow irrigation of cotton. Journal of Irrigation and Drainage Engineering. 2018;**144**(5):04018005

[20] Al-Kaisi MM, Broner I. Crop Water Use and Growth Stages. Fort Collins, CO, USA: Colorado State University Extension; 2009

[21] Lakhiar IA, Yan H, Zhang C, Wang G, He B, Hao B, et al. A review of precision irrigation water-saving technology under changing climate for enhancing water use efficiency, crop yield, and environmental footprints. Agriculture. 2024;**14**(7):1141

[22] Alharbi S, Felemban A, Abdelrahim A, Al-Dakhil M. Agricultural and technology-based strategies to improve water-use efficiency in arid and semiarid areas. Water. 2024;**16**(13):1842

[23] Orazgylyjov B, Dowletov H, Balyshev T, Rejepov E, Hamydov R. The use of smart systems in cotton irrigation as a way of water saving in the climatic conditions of the Dashoguz province. In: BIO web of conferences. Vol. 71. EDP Sciences; 2023. p. 01091

[24] Ya G, Romashchenko M, Kovalchuk V, Matiash T, Voitovich O. Using smart technologies in irrigation management. In: International Commission on Irrigation and Drainage: 3rd World Irrigation Forum (WIF3). Bali, Indonesia: WIF3; 2019

[25] Imran MA, Ali A, Ashfaq M, Hassan S, Culas R, Ma C. Impact of climate smart agriculture (CSA) through sustainable irrigation management on resource use efficiency: A sustainable production alternative for cotton. Land Use Policy. 2019;**88**:104113

[26] Dhir A, Pal RK, Kingra PK, Kaur R. Climate-smart cotton (Gossypium herbaceum) crop production in Punjab: A comprehensive review of sustainable management practices. The Indian Journal of Agricultural Sciences. 2024;**94**(2):119-128

[27] Hussain S, Ahmad A, Wajid A, Khaliq T, Hussain N, Mubeen M, et al. Irrigation scheduling for cotton cultivation. In: Cotton Production and Uses: Agronomy, Crop Protection, and Postharvest Technologies. Singapore: Springer Singapore; 2020. pp. 59-80

[28] Çetin Ö. Irrigation management and innovative approaches in cotton under climate change. In: Agriculture and Water Management Under Climate change. Cham: Springer Nature Switzerland; 2025. pp. 65-92

[29] Foley J, Gillies M, McCarthy A, Kelderman S, Eberhard J, Shippam R. Smarter Irrigation for Profit 2 RRDP2002 Precise Real-Time Automated Cotton and Dairy Irrigation for Improved Water Productivity. Cotton Research and Development Corporation; 2022. pp. 1-150

[30] Gamal Y, Soltan A, Said LA, Madian AH, Radwan AG. Smart

irrigation systems: Overview. IEEE Access. 2023

[31] Pandiyarajan P. Review of smart agricultural technologies for precision pest control and sustainable water in cotton farming. In: 2025 International Conference on Inventive Computation Technologies (ICICT). IEEE Access PP(99):1-1. Australia: University of Southern Queensland; 2025. pp. 1799-1809. DOI: 10.1109/ACCESS.2023.3251655

[32] Calderan Gregolin A, Contreras M, Salgado-Funes E, Tejada G, Zabaleta I. The use of efficient smart agro 4.0 irrigation systems for cotton in Peru and its public policy implications. World. Water Policy. 2023;**9**(4):746-755

[33] McCarthy A, Foley J, Jamali H, Raedts P, Hills J. Machine vision tools for smarter irrigation of dairy pasture and cotton. In: 2022 Irrigation Australia Conference and Exhibition. Australia: University of Southern Queensland; 2022. p. 19

[34] Yang ZY, Xia WK, Chu HQ, Su WH, Wang RF, Wang H. A comprehensive review of deep learning applications in cotton industry: From field monitoring to smart processing. Plants. 2025;**14**(10):1481

[35] Wang Y, Yang Y. Research on application of smart agriculture in cotton production management. In: 2020 International Workshop on Electronic Communication and Artificial Intelligence (IWECAI). Piscataway, New Jersey, USA: IEEE; 2020. pp. 120-123

[36] Rasheed MA, Mohammed AM. Infrastructure for smart agriculture for cotton crops based on the internet of things. In: IOP Conference Series: Earth and Environmental Science. Vol. 761. Bristol, United Kingdom: IOP Publishing Ltd.; 2021. p. 012018

[37] Zhang S, Jing H, Dong J, Su Y, Hu Z, Bao L, et al. Accurate estimation of plant water content in cotton using UAV multi-source and multi-stage data. Drones. 2025;**9**(3):163

[38] Du C, Zhang L, Ma X, Lou X, Shan Y, Li H, et al. A cotton high-efficiency water-fertilizer control system using wireless sensor network for precision agriculture. PRO. 2021;**9**(10):1693

[39] Kakkavou K, Gemtou M, Fountas S. Drivers and barriers to the adoption of precision irrigation technologies in olive and cotton farming—Lessons from Messenia and Thessaly regions in Greece. Smart Agricultural Technology. 2024;**7**:100401

[40] Shao L, Gong J, Fan W, Zhang Z, Zhang M. Cost comparison between digital management and traditional management of cotton fields—Evidence from cotton fields in Xinjiang, China. Agriculture. 2022;**12**(8):1105

[41] Tata Power Solar. Pilot project on solar drip irrigation for cotton farmers in Maharashtra. 2021. Available from: https://www.tatapowersolar.com

[42] USDA-ARS. IrrigatorPro—Decision Support Tool for Cotton Irrigation. Agricultural Research Service, United States Department of Agriculture; 2020. Available from: https://www.ars.usda.gov

[43] Cotton Research and Development Corporation (CRDC). Smart irrigation technologies in Australian cotton farms. 2021. Available from: https://www.crdc.com.au

[44] Raihan A. A review of recent advances, challenges, and potential

future directions of climate-smart agriculture. In: Proceedings of the International Conference on Climate-Smart Agriculture. Bhubaneswar, Odisha, India: Siksha 'O' Anusandhan University; 2024

[45] Qian M, Qian C, Xu G, Tian P, Yu W. Smart irrigation systems from cyber–physical perspective: State of art and future directions. Future Internet. 2024;**16**(7):234

[46] Balyan S, Jangir H, Tripathi SN, Tripathi A, Jhang T, Pandey P. Seeding a sustainable future: Navigating the digital horizon of smart agriculture. Sustainability. 2024;**16**(2):475

[47] Vaddula Y, Singh K. Progression of drip irrigation and fertigation in cotton across the globe and its future perspectives for sustainable agriculture: An overview. Applied Water Science. 2023;**13**(9):177

[48] Tausif M, Jabbar A, Naeem MS, Basit A, Ahmad F, Cassidy T. Cotton in the new millennium: Advances, economics, perceptions and problems. Textile Progress. 2018;**50**(1):1-66

[49] Aarif KOM, Alam A, Hotak Y. Smart sensor technologies shaping the future of precision agriculture: Recent advances and future outlooks. Journal of Sensors. 2025;**1**:2460098

[50] Lawal Y. AI and IoT-Based Smart Irrigation: A Review of Challenges and Future Trends. Ankara, Turkey: Turkish Science and Technology Publishing; 2025. pp. 770-776

[51] Harshitha AG, Kumar S, Jain A. A review on organic cotton: Various challenges, issues and application for SMART agriculture. In: 2021 10th International Conference on System Modeling & Advancement in Research Trends (SMART). Piscataway, New Jersey, USA: IEEE; 2021. pp. 143-149

[52] Zhao J, Liu D, Huang R. A review of climate-smart agriculture: Recent advancements, challenges, and future directions. Sustainability. 2023;**15**(4):3404

[53] Anjum MN, Cheema MJ, Hussain F, Wu RS. Precision irrigation: Challenges and opportunities. Precision Agriculture. 2023;**1**:85-101

Chapter 4

Digital Innovations in Cotton Farming

Ghulam Sarwar, Muazzama Batool and Muhammad Tauseef

Abstract

Challenges to cotton farming in recent scenario of climatic changes can be mitigated by digital innovations and applications in cotton farming. Digital innovations can offer numerous opportunities for sustainability and improved productivity. Technologies including remote sensing, digital information system, and artificial intelligence can boost/redefine crop monitoring and management practices. Implementation of technologies like smart sensors, satellite imaging, management software/machine learning, autonomous machinery can redefine decision making about crop with the higher yield and lower costs of inputs. Environmental impacts of hazardous crop protection chemicals and inefficient use of irrigation water contributes to loss in biodiversity. In a market chain, digital innovations can assist in transparency and traceability to meet consumer demand and sustainability.

Keywords: precision tools, apps in agriculture, drones, IoT, sensors, productivity, technology in agriculture

1. Introduction

Cotton is widely cultivated in warm climates around the globe to obtain a natural, fluffy, and soft fiber which is spun to make textiles [1]. It is most important non-food crop of many countries, which contributes to economy and industry. China, India, USA, Egypt, Uzbekistan, and Pakistan are the major cotton-producing countries [2]. Cotton plant is a perennial shrub but cultivated as annual crop for efficient farming practices and pest control. The fluffy fiber grows in a boll, which is a protective case around the seeds of cotton. The fiber is pure cellulose and can contain traces of waxes, fats, pectin, and water [3].

Cotton belongs to genus Gossypium of family Malvaceae which have almost 50 species out of which 4 are reported to be domesticated worldwide since ancient times. *Gossypium herbaceum* and *G. arborerum* are native to South Africa and Indian subcontinent, respectively, and contribute to 2% global cultivation. *Gossypium barbadense* is native to South America and contributes to around 5% of total cotton cultivation having extra-long staple. *Gossypium hirsutum* (native to central America) is commonly known as upland cotton/American cotton and cultivated on more than 90% area around the globe [4]. The farming processes of cotton involve several practices including land preparation, sowing, irrigation, pest management, harvesting, and post-harvest processing. While its fiber is a key raw material for textile industry, its seeds are used to make oil and cotton seed cake, which is a food source for livestock [5].

1.1 Historical overview of cotton cultivation

The word "cotton" is derived from an Arabic word *qutun or qutn*. The oldest record of cotton fabric (thought to be spun from *G. barbadense*) has been found in Peru which dates back to 6000 BC which evidences that *G. barbadense* was domesticated the earliest. Oldest cotton bolls (approximately 5500 BC) were found in a cave in Mexico. Domestication of *G. hirsutum* in Mexico is recorded between 3400 and 2300 BC and people of Santiago grew, spun, wove, and sewed cotton themselves. In 4200 BC *G. barbadense* was the backbone of coastal line of Peru. The people were using cotton woven clothes in sixteenth century in Mexico and Peru [6].

The archaeological discovery puts the early cotton cultivation record around 6000 BC, in Mehrgarh near Indus Civilization of Pakistan. The cotton fabric fragments were also reported in Mohenjo-Daro and Harappa dates 2500 BC, which suggests its importance as a trade commodity. Cotton pollen has been reported from Balakot, seeds from Banawali (1900 BC), Sangol (1400 BC), Kacchh (1700 BC), and Gorakhpur (800 BC), and fibers at Sringaverapura in 700 BC [7].

British Trading Empire during sixteenth century raised cotton importance in Europe and cotton woven fabrics like calico and chintz became popular during Cultural transformations in Europe. During the regime of East India Company, cotton was imported into Britain. Indian cotton industry gained competitive advantage for import of cotton and fabrics to Britain but later on East India Company invested in labor saving technical processes and tariffs to ban on Indian trade. The British cotton industry developed till nineteenth century and became a monopoly. Later on, the cotton trade and economy became a complex sector with multiple global interconnections. Multiple nations seek opportunities in this sector but managing market volatility, social inequalities, and sustainable initiatives for better cultivation are crucial for flourishing the industry [8].

1.2 Global and regional importance of cotton

Due to its wide range of applications, value-added products, economic and social benefits to countries particularly to developing nations, cotton has attained the status of significant crop. It is an important contributor to GDP and provides livelihoods to millions of people in developing countries. More than 30% of world's fiber comes from cotton which is used in apparel, home furnishing, and various medical supplies. China, India, and USA are the largest consumers as well as the producers of cotton. China and USA have large domestic textile industry, advanced mechanized farming, and account for major global cotton exports. Brazil is the new emerging cotton industry, which has vastly invested in technology driven farming and market hunting (**Table 1**).

1.3 Challenges in traditional cotton farming

Today agricultural landscape is being highly affected by several challenges attributed by biotic and abiotic factors. These challenges turn out to be a risk factor by outdated farming practices, substandard or adulterated seed, and climatic changes leading to increased production inputs and lower yields [9]. Sensitivity of cotton to

Rank	Name of Country	Production (million metric tons)	% of global production
1	China	5.9	24
2	India	5.6	23
3	Brazil,	3.1	13
4	USA	2.6	11
5	Pakistan	1.51	6
6	Australia	1	4
7	Turkey	0.69	3
8	Uzbekistan	0.63	3
9	Argentina	0.35	1
10	Mali	0.28	1

Table 1.
Largest cotton-producing countries in 2023/2024. Source: USDA.

abiotic stresses (heat, drought, and salinity) requires smart and efficient management practices. In water-scarce areas, pumping of groundwater for irrigation may worsen the scenarios. Inefficient irrigation practices that leads to water logging and salinity degrade soil, which are exacerbated monoculture practices that decrease soil fertility. Crops grown in poor soil are attributed by poor health, which are vulnerable to pest infestation and hence increase financial pressure on farmer [10, 11].

1.4 Role of innovation in cotton farming

Optimizing the best technical advancements in a geographical area is crucial for ensuring productivity and sustainability. It involves innovating and adopting new technologies, practices, methods, and approaches to improve ecosystem elements through precision agriculture, mechanization, and biotechnology [12, 13]. Precision agriculture techniques are useful for efficient decision making. Mechanization application in cotton pest management and irrigation can be useful in resource efficiency and attaining global food security [14, 15].

Although the digital tools in agriculture are reported to have multiple benefits for farming community from planning a crop to market but in developing regions of the world there are many barriers to adopt recent technologies and attain a sustainable production system [16]. Important key areas of digital tools application in agriculture are defined in **Figure 1**.

Outcomes of technology application has been differentiated as:

- Economic outcomes
- Increase in income and productivity
- Market efficiency and transparency
- Cost effectiveness
- Increased per capita

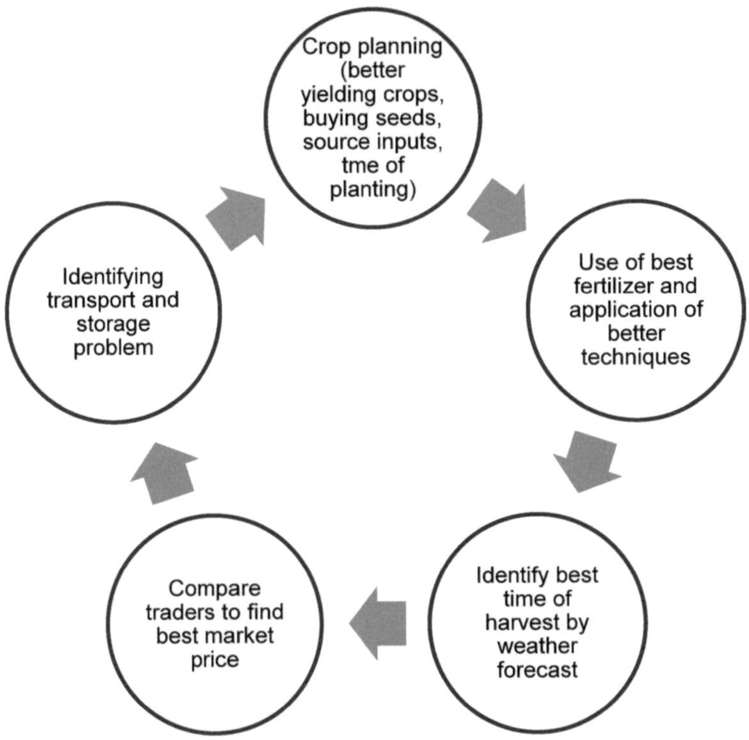

Figure 1.
Highlights of digital tools in agricultural ecosystems.

- Improved livelihoods
- Social outcomes
- Freedom of decision making
- Improved household nutrition
- Environmental outcomes
- Climate resilience
- Environmental sustainability

2. The digital agriculture ecosystem

2.1 What is digital agriculture?

Application of digitization through advanced tools to optimize agricultural technologies and farming is called digital agriculture or smart farming of e-agriculture [17]. It involves data collection, analysis, and implementation through certain satellites, drones, sensors, and other equipment for efficiency, productivity, and sustainability. The FAO of united nations has designated digitization processes as *digital agricultural revolution*. Digital agriculture can help accelerate the agenda of Sustainable Development Goals by the United Nations by providing the real time information about farms and resource use efficiency [18].

2.2 Internet of Things (IoT)

The Internet of Things or IoT is integration of people, technology, and processes with interconnected devices or sensors for monitoring and evaluation. The term was first coined by Kevin Ashton of Proctor and Gamble in 1999 [19]. Internet of Things in agriculture refers to the use of small interconnected devices which record, exchange, and analyze real time data. These devices also monitor environmental conditions, can start automated tasks, and help in data-driven efficient decision making [20, 21].

Key applications of IoT in agriculture include crop and soil monitoring sensors, which measure moisture, temperature, nutrient levels, and chlorophyll levels. Data driven through these sensors is analyzed and helpful in irrigation and fertilizer application schedules. On-farm tracking of temperature, wind speed, humidity, and rainfall through smart devices can help manipulate irrigation, saving water, and increasing crop efficiency. This data exchange can also help in sowing plans, crop protection, chemical applications, and harvesting [22].

Sensors and controllers in greenhouse grown crops manage light, temperature, humidity, and CO_2 levels minimizing manual inputs to optimize indoor conditions. IoT-based drones and soft wares forecast and detect early infestation, prior to ETLs, which implies wise application of pesticides and reducing the risk of development of insecticide resistance [23].

Artificial intelligence approaches transform agroecosystems by helping in resource management, reducing waste, agribusinesses, and climatic adaptations resulting in enhanced productivity. Artificial intelligence can assist farmers in crop monitoring, infestation detection, targeted application, yield predictions, autonomous use of machinery, assisted mechanization practices, modeling of localized impact of any factor, and market hunting for better profitability [24].

Geographic information system (GIS) is another technique for efficient decision making about farming which refers to mapping tools using spatial data for management of farm and environmental information. These location-based decisions generated by the tools are helpful in the assessment of land suitability to a certain crop, soil topography mapping, crop growth and health monitoring with specific interest on stress, water availability, drainage, and disaster management [25].

Digital innovations in farming also include development of *mobile applications* for modern farmers which offer monitoring and decision making to enhance productivity. Although mobile applications are developed according to different aspects of farm practices, they can contribute greatly to efficient decision making. Mobile apps can also provide information about financial services and market access. Farm management apps are helpful in planting operations, inputs application, and harvesting. Examples of farm management apps are *AgriWebb*, *Cropln*, *Farmlogs.* Climate and weather forecasting apps are designed to provide localized weather forecasts to plant cropping activities. Examples include *The Weather Channel, Agrible, Climacell.* Soil and crop health and irrigation management apps include *Plantix*, *CropWise*, *NetBeat.* Other apps include *KisanSuvidha* (for market decision guide), *IFFCO* (helping farmers to make informed decisions), and *Agri.360.io* (platform to manage farm operations) [26, 27]. A valuable tool for cotton farmers offering diagnostics and information about production is designed by ICAC named as ICAC Cotton Expert. It assists farmers with multilingual support, offline advisory, and also process pictorial and voice data by farmers.

2.3 Benefits and barriers to adopt digital tools in developing regions

Digital innovation applications in agriculture have resulted in sustainable management and production system in many regions of world. Africa and India are the recent examples of developing regions implying techniques to reform agriculture where the outcome and impact are found to reshape the traditional trends and life [16]. Barriers to adopt the technologies in the developing countries are hindered by multiple factors. High costs in initial investments for digitalization, data infrastructure, hardware, and software are the main hurdles for small land holdings in developing regions. Lack of internet facilities and electricity in remote areas is another pitfall to adopt the modern technologies. Countries with having low literacy rates are more prone to poor implementation of regulations and infrastructure where people do not trust new technologies and prefer traditional practices and cultural norms. Moreover, complexity of some digital tools makes it more difficult for illiterate farmers to adopt and operate [28].

2.4 Digital transformation trends in agribusiness

Implementation of digital technologies is complementary to attain food and fiber security. Digital transformations can not only redefine production systems but can help identify inefficiencies in supply chains and facilitation of redistribution of surplus food to required areas. Technologies and artificial intelligence can prevent and predict food spoilage and hence reduce food waste. Transformations in the agribusiness are driven by increased demand of efficiency, transparency, sustainability, and value chain management. These targets can be achieved by blockchain, robotics, autonomous machinery, and artificial intelligence coupled and IoT. E-commerce platforms like AgroStar (India) and TwigsFoods (Africa) have enabled farmers to sell their products through mobile application and also by surpassing middle man [29, 30].

3. Precision agriculture in cotton farming

Precision agriculture in cotton refers to the application of advanced technologies in cotton farming such as sensors, GPS, drones, algorithms, AI, and other digital tools to reduce input cost and manage cotton fields more efficiently for increased yield and fiber quality.

3.1 Role of precision agriculture in cotton

Precision agriculture plays critical role at each stage of cotton crop, that is, from planning to harvesting by optimizing input use, reducing environmental impact and improving yield. Precise management of resources is achieved through collecting and analyzing data to make wise and site-specific decisions [31]. Role of precision agriculture in cotton farming is summarized in **Table 2**.

Role	Technology adopted	Outcome	Long-term benefits
Precision planting	GPS guided planters and seed drills	Accurate plant spacing and seed placement	Uniform plant population, crop stand and improved yield
Site-specific fertilizer application	Soil analysis sensors, GIS, VRT	Nutrient application at required place	Reduced runoff, better soil health, cost efficient
Efficient irrigation	Soil sensors, Weather data	Optimal water application	Increased water use efficiency
Pest monitoring	Drones, AI-based recognitions, imaging	Early detection of infestations	Wise application of chemicals, reduced input cost, healthier crop
Crop health assessment	Drones, remote sensing,	Detection of plant vigor and growth anomalies	Optimized crop management decisions
Harvest management	Yield detectors, GPD-coupled harvesters	Yield data collection per field basis	Low- and high-yielding areas identification help planning for the next crop
Weather responsive decisions	Weather forecasting tools	Efficient decisions on irrigations, spraying, harvesting	Lower risk of adverse weather events
Weed management	Automated sprayers	Targets only weed infested areas	Saves input costs and environmental risks
Traceability and compliance	Digital record keeping, blockchain	Track crop journey at each stage	Build consumer trust, enable meeting export standards and certifications

Table 2.
Summary of the role of precision agriculture.

3.2 GPS and remote sensing for land mapping

For uniformity and smooth input application, land preparation is the first step for any crop. A comprehensive overview of earth surface and precise spatial information is necessary for land use planning and cultivated crop monitoring. Remote sensing and GPS are the powerful tools for the purpose. Remote sensing collects the data about land features from a distance by using ground control points (accurate positions) referenced by the GPS. The combination of both techniques provides an accurate mapping of land cover and land resources. GPS is also helpful to point crop boundary mapping to create digital maps for crop management and planning. Other implications of GPS are now being extensively reported in farm mechanics where GPS-guided machinery (tractors and sprayers) ensures precise sowing, soil sampling, irrigation, fertilizer, and pesticide applications. Use of GPS-guided farm mechanics have increased efficiency of inputs and reduced overlapping or repeating [32].

Remote sensing analyzes ground control points of GPS to map and identify different land covers like forests, water bodies, field covers, and urban areas. Images obtained from drones, aircrafts, and satellites (multispectral, hyperspectral, thermal, and time series imagery) can distinguish cotton crop from other crops. Crop health, growth, flowering, pest infestation, abiotic stress, and yield estimation can be analyzed through remote sensing by using various techniques like normalized vegetation index (NDVI), algorithms (vegetation index and biomass). These analyses can also be helpful to plan mechanization practices in crop, irrigation, and harvest [33].

3.3 Variable rate technology for input application in cotton

Variable rate technology is defined as application of inputs (fertilizers, pesticides, seed, and irrigation) at different rates across the farm, based on specific needs. It can improve input efficiency, reduce waste, and potentially increase yield and profit in cotton production. Variable rate technology works by collecting data about field variability like soil sampling, satellite imaging, drone imaging, and software. These images are then used to create maps to define deficiency sites for optimal application rates at different areas of field. Reduced and specific application of inputs leads to lower costs per unit of production along with the reduced environmental impacts of hazardous chemicals. Examples of some variable rate applications in cotton refer to fertilizer application, variable rate seeding, variable rate pesticide application, and irrigation [34, 35].

3.4 Yield mapping and data-driven decision making

Prior and during harvest, georeferenced data based on crop health, yield, and moisture content like parameters are recorded throughout the farm and analyzed to illustrate various spatial maps to prescribe variation in yield. Yield mapping system works through a *Yield Monitor*, which is accompanied with tools such as GPS receiver, moisture sensor, data logger, and software. Benefits of yield mapping tools include optimized input use, higher yields, and sustainability. Combined with NDVI, soil sampling, and weather forecasting, it can enable the farmers with the smarter management decisions to boost productivity and profitability [36, 37].

4. IoT applications in cotton cultivation

4.1 IoT-enabled devices: sensors, drones, and automated machines

IoT can open a new era of crop yield and sustainability through optimized input which is already discussed in above sections [38]. Application of IoT tools for cotton is summarized in **Table 3**.

5. Artificial intelligence and machine learning

Artificial intelligence in agriculture is the use of data analytics and machine learning for profitability, efficiency, and sustainability across the entire chain of production to value addition. It is attributed by real-time monitoring of biotic and abiotic factors, farm mechanization operations, and textile manufacturing. Machine learning in agriculture is the application of algorithms to analyze data obtained from various digital sources for crop improvement and productivity. Some examples of AI powered solution for cotton cultivation include BASF and John Deere. Combination of both AI and ML can contribute cotton industry through:

- AI powered drones and satellite imaging to monitor crop health, pest infestation, and growth assessment. ML can analyze these algorithms and provide efficient decisions enabling timely interventions

- Crop, soil, and weather data be analyzed for optimized input use and yield prediction

- Computer vision systems can perceive weeds and pests enabling smart application of crop protection chemicals with the reduced environmental effect

- AI-guided autonomous cotton pickers/robots can perform efficient picking which reduces labor dependency

- ML-guided spectroscopy and machine vision can evaluate fiber quality after harvesting

- Carbon foot prints, water usage, and soil health can also be monitored through AI.

IoT applications in cotton farming			
	Devices	**Functions**	**Benefits**
Smart soil monitoring	Soil sensors Temperature sensors EC sensors	Real time tracking of soil conditions	Optimized input application
Weather stations	On filed IoT weather stations	Collect micro-climate data	Planning for sowing, spraying, harvesting, etc.
Precision irrigation system	IoT connected sprinkler or drip	Auto-regulated water delivery	Water conservation
Health monitoring	Drones NDVI cameras Cloud database	Assessment of plant health and stress	Detection and targeted based action
Smart cotton harvesting	Yield monitor cotton flow sensors GPS-enabled harvesters	Real time yield mapping and monitoring	Accurate yield data for profitability analysis
Logistics and post-harvest monitoring	RFID tags GPS trackers Temperature and humidity sensors during storage	Tracking of bales from filed to gin, and quality monitoring	Post-harvest losses are minimized
Centralized farm management Platforms	Mobile-based dashboards having all data of sensors	Real times alerts, historical analysis, automated controls	Single click controls of all filed operations
Pest and disease prediction	Insect traps with cameras and sensors Pheromone traps AI-backed image and analysis	Detection of population trends and disease onset	supports timely application of crop protection measure

Table 3.
Application of IoT in cotton farming.

6. Digital tools for pest management

In the modern era of agriculture pests (insect pests and diseases) can be monitored and managed through digital technologies like remote sensing, AI, and IoT. Because of early detection, forecasting, and precise decision guidance, these tools have become popular to attribute IPM strategy against any pest. Key digital tools and technologies employed for monitoring and management of pests are described in the table below (**Table 4**).

Management of pests through digital tools			
	Tools	**Use**	**Benefits**
Remote sensing and satellite imaging	NDVI Multispectral cameras Hyperspectral cameras	Monitoring of crop health and identification of pest outbreaks	Early detection of pests prior to damage
IoT-based pest monitoring system	Trapview Pessl Intstruments Semios	Automated traps with sensors to detect and count pests	Alert for thresholds
Mobile apps	PlantiX AgriApp eLocust3	Identification, diagnosis, and treatment	Real time decision making
GIS	GPS Remote sensors Software	Spatial mapping of infestation	Targeted control
AI and ML	PlantVillageNuru	Prediction, identification, models for outbreaks	AI-based recognition and identification for decision making
Decision support system	IPMwise Cabi's Pest Smart	Combine weather and crop data for pest management options	IPM planning
Predictive analytics and forecasting tools	Climex like phenology models for insect pests	Model pest life cycle for its forecasted outbreak	Preventive action before infestation
Drones and UAVs	Fixed wing Multirotor VTOL	Aerial scouting and spraying	Efficient use of production and protection chemicals
Robotics and automation	EcoRobotics BlueRive's See Spray	Mechanical pest removal, precision weeding, and spraying	Reduced input and labor cost
Blockchain and data platforms	AgUnity IBM Food Trust	Track pest control input and traceability in pest management	Transparency in pest management practices

Table 4.
Management of pests through digital tools.

7. Supply chain traceability and blockchain

7.1 Importance of transparency in cotton value chain

Nowadays cotton consumers are increasingly conscious of where and how products are made. Transparency about sourcing, labor conditions, and environmental impact builds brand credibility and loyalty. Transparency in the cotton value chain is crucial for building consumer trust, ensuring ethical labor and environmental practices, improving supply chain efficiency, meeting certification standards, and attracting sustainable investments. It enables traceability from farm to fashion, supporting responsible and sustainable cotton production.

7.2 Blockchain for tracking cotton from farm to fabric

Mobile apps allow farmers to use tools for crop data review, find current market prices, and pay for services or goods. Each of these gadgets get such data, which can be transferred to the blockchain to keep everything up front. If something odd is detected by AI in the data or blockchain, it blocks it. Because of this digital system, small-holding farmers can benefit from being involved in the value chain. With the help of QR code or NFC tags with block chain, everyone from farmers to fashion brands may look up the process and transactions behind the clothes they wear or sell. Even more, digital innovation makes it possible to save and use records in real time, avoiding the costs and efforts of building infrastructure and hiring staff.

Using blockchain, it becomes possible to have secure, open and unchangeable tracking of how cotton is made, supporting both authenticity and ethical purchasing and being kind to the environment [39]. Soil condition, the amount and types of pesticides used, and the time of harvest can be monitored using sensors, drones, and GPS devices [40].

7.3 Digital tagging and QR-based verification

Digital tagging and QR codes have become very important for making the cotton supply chain more transparent and traceable. If each batch or product has a special QR code, stakeholders can quickly look up details about its origin, how it was processed. This method not only makes sure customers trust what the product claims but also ensures that supply chain is held accountable. A variety of studies suggest that digital tagging and blockchain combined help ensure no fraud, confirm production standards, and smoothly watch over ethical practices and environmental impact in cotton manufacturing [41, 42].

The arrival of digital tagging and QR code verification has increased the openness, traceability, and accuracy of cotton products at every stage of production. Typically, cotton's supply chain is challenged by not being able to fully see operations, fighting counterfeit products and checking if claims about the environmental impact are real. Digital tags using QR codes can be used for each batch of cotton or finished textiles so that every item can be linked to secure, detailed information on online platforms, often using blockchain storage [41, 42].

7.4 Improving trust among buyers and producers

A strong bond of trust between the people who grow and purchase cotton is needed for the supply chain to function well and last. Issues such as information gaps and doubts about quality affect how strong relationships can be. Thanks to digital improvements, it is now possible to track where products come from, their quality, and their compliance with social and environmental rules, which helps to prevent conflicts and fraud [41, 42]. Mobile technology and online market spaces make it easier for smallholders to receive fair-priced purchases, market updates, and quick payments, which benefits their economy and attracts more buyers [43]. Uses of QR technology for checking certifications have shown that it makes people more confident and readier to spend more on products [41, 42], because of these tools, there is more transparency, tracking of cotton products, and open communications which closes trust gaps and encourages fairness and resiliency worldwide.

8. Mobile applications and advisory platforms

8.1 Farmer-friendly apps for decision support

Apps that support farmers have changed farming by giving instant details on crops, pest problems, watering, weather, and market trends, boosting how much and how sustainably farmers can produce [41–43], because these apps focus on accessibility, and users can set them up in several languages, collect little data, and control IoT sensors, which allows them to get important information and insights that relate to what is happening in the field [44]. It is confirmed by research that such apps increase crop production, use less input for farming, and encourage eco-friendly actions by regulating fertilizers, pests, and irrigation [22, 36, 45]. Digital innovations in marketing and finance also help farmers by giving them stronger power in their supply chains [46]. Examples are Plantix and CropIn which help farmers, extension workers, and experts to share information and support the growth of a durable and savvy agriculture field [47, 48].

8.2 SMS-based advisory in rural areas

Real-time farming advice is now given to rural people through SMS, as not all communities have the internet and smartphones [49]. With the help of these services, farmers are informed about weather, pest problems, crops management, irrigation, and market prices, helping them increase how much they produce and earn [50]. Cole and Fernando point out that SMS evenly distributes updates across regions, partly compensating for a lack of personnel in extension programs [51].

Adding IVR and regional content to SMS notices boosts how successful they are, making them more available for people with low-reading skills [52]. With tools like mKisan and Esoko, India and Ghana have seen big improvements in their harvests, control of pests, and ways to sell their crops [47, 53]. Even so, there are areas where improvement is needed, especially relevance of the content, trust in the network, and market adoption which require these aspects to be solved by partnering NGOs, governments, and telecommunication providers [54]. Marrying SMS systems with radio, on-field events, and farmer groups allows these advisories to have a better impact [55]. All in all, these solutions that are both low-cost and scalable support food security, eco-friendly farming, and agricultural jobs in rural areas [56].

8.3 E-extension services and knowledge dissemination

Digital tools such as apps, websites, SMS, social media, IVR, and video tutorials are part of e-extension and help in giving farmers accurate advice fast, even if there are distance, expense, and personnel issues [57]. Using ICTs, these services support access to knowledge on farming, pest control, soil condition, and climate, which leads to making decisions based on data [58]. Interactive tools, like online forums and Expert chatbots, allow farmers to get immediate help and advice from other farmers [59].

A number of researchers have found that using e-extension increases the uptake of new techniques, improves farm success, and assists smallholder farmers by giving them extra resources, especially when emergencies like pest invasions or severe weather appear [42]. By mixing media channels like SMS, radio, and mobile apps, a larger number of people can benefit, no matter their education or access to technology [56]. On the other hand, there are issues like users' digital skills, limitations to access in some parts, adapting to unique needs, and maintaining interest [54]. Developing and applying relevant solutions needs governments, NGOs, telecom providers, and farmers to cooperate closely in the design process [60]. When e-extension is used properly, it helps connect agricultural research, advisory services, and farmers at the field level so they can adapt to the increasing use of technology in farming [61].

8.4 Examples of digital platforms in South Asia and Africa

Digital farming technology has started running in South Asia and Africa, giving more advice and updates to farmers with small farms. Farmers in India get information such as weather updates, advice on pests, tips for farming, and market prices from the government *via* mKisan, which sends out information through SMS and apps in many regional languages. A lot of farmers depend on this platform to access important farming advice quickly [58].

Using locally recorded videos on mobile devices, similar to that of Digital Green, India and several other African countries are able to communicate about better ways to farm. Videos in local languages shared by Digital Green help farmers easily discover new ideas and start using better farming techniques [50]. In Africa, using Esoko in Ghana provides farmers with updates on the market, warnings about weather, and access to possible buyers and sellers through SMS, helping them find good prices. In Kenya, iCow is a leading example because it gives SMS advice on animal care and farming, assisting farmers and increasing their harvest [47].

Through the Kisan Suvidha app, the government collects multiple advisory services concerning weather, pests, and markets and provide this information in farmers' native languages. By designing ICT to match local situations and the people who use them, information barriers are overcome, and farming grows more sustainable and life in rural regions improves across South Asia and Africa.

9. Data analytics for farm optimization

9.1 Collecting and managing farm data

Data collected by precision agriculture is improving cotton farming. Monitoring soil, microclimates, and pest populations is possible using IoT sensors, drones, and GPS, which reduces input waste by 15–20% [43]. Information from sensors, weather predictions, and the market are combined in integrated cloud platforms to help with planting schedules and pest solutions [62]. Because of blockchain technology, it is now easier for people to see the supply chain, as 78% of farmers producing fair-trade cotton use verification online [47]. Despite issues with how accurate data is and digital skills, these tools support small farmers.

- Improve farm productivity by 25–30%.
- Use less water and fertilizer.
- Focus on sustainability certification standards.

9.2 Predictive analytics for input efficiency

Farmers now use predictive analytics to assess their data, plus machine learning, statistics, and big data together which leads to cutting waste, raising crop quality and better managing fertilizers, water, and pesticides [43, 62]. It helps to estimate when pest problems or diseases may occur, allowing immediate action and reduced expenses. Following weather and soil conditions helps farmers use less water, which is necessary because cotton requires a lot of water. Using IoT and mobile technologies, it assists with different agricultural issues in a sustainable, precise way [63–65]. Since there are issues with data quality, model trustworthiness, and farmer acceptance, more research and training should be carried out [66, 67].

9.3 Decision Support Systems (DSS) for farm planning

Decision Support Systems (DSS) in cotton farming collect data on soil, weather, pests, and markets to provide tailored advice on planting, irrigation, fertilization, and pest control. Using automation, forecasting, and computations, DSS optimize outcomes while reducing costs. By delivering precise, field-specific guidance, DSS help farmers enhance productivity and environmental sustainability. Integration with mobile apps and IoT devices further simplifies and improves farm management [43, 62].

9.4 Cloud computing and data dashboards for farmers

Cloud computing has greatly simplified how farmers look after, access, and secure their large amounts of data. Through online devices, cloud platforms supply farmers with current weather, soil conditions, market pricing, and readings from remote sensors. By relying on cloud-based dashboards, farmers and stakeholders can easily see crop information, resource details, and farming changes, which helps them make quicker choices and manage resources better. With the help of mobile apps, smallholder farmers in remote places can also use advanced technologies normally available in cities [43, 62].

10. Economic and environmental impact

10.1 Cost-benefit analysis of digital interventions

A cost-benefit analysis for tools in digital agriculture such as mobile advisory, precision farming, blockchain technologies, and decision systems, checks if they are worthwhile and effective. While it may cost money to set up and teach others, these tools often yield a lot of rewards in the form of productivity, less waste, broader market reach, and transparency in the supply chain. Using mobile advisories, farmers can avoid wasting fertilizer and water and get better yields. Even if blockchain and IoT need a significant investment at first, the payoff is reliable traceability and less fraud in the future. If they are designed for each community and bolstered by skills training, digital solutions usually give businesses and economies more benefits than they cost [43, 58, 68] .

10.2 Environmental sustainability through optimized inputs

Cotton can be grown sustainably by cutting back on water, fertilizers, and pesticides. Using too much water and chemicals in cotton farming might lead to damaged soil, polluted water, and biodiversity problems. The use of sensors, weather information, and crop observations leads to precise input timing and placement, thanks to precision application [67]. Using both precision agriculture and predictive analytics leads to less need for over-irrigation, which helps prevent both water waste and soil salinization. Targeted use of fertilizers and pesticides on specific plots helps reduce pollution and harmful chemicals and helps maintain a balance in the environment, also increasing the quality of crop yield and fiber [62]. By using digital tools, ongoing monitoring, and IoT systems, it is possible to manage cotton production in ways that fit with internationally agreed sustainability goals. Rational use of input materials ensures the farming of cotton does not harm the environment and offers a sustainable future Impact on Cotton Yield and Quality [63].

Taking care of precise input use such as water, fertilizer, and pest control, helps growers receive high yields and better quality cotton fibers. Using precision agriculture, interventions can be applied promptly and based on data, resulting in safer plants, better health, and bigger yields [43]. Managing soil nutrients carefully helps bolls grow well and provides strong fibers and making sure that water is not wasted by over-irrigation prevents harm to the fibers. Controlling pests and diseases helps preserve healthy plants and clean fibers and this leads to better fiber length, fineness, and homogeneity, important for textile production [62]. Monitoring and managing cotton in real time, with the help of data analytics and digital equipment, leads to stable good quality cotton. Because of these improvements, cultivation of cotton yields higher productivity and the market is supplied with the high-quality cotton fiber [47].

10.3 Socioeconomic benefits to smallholder farmers

New advances and eco-friendly ways of farming cotton profit smallholder farmers socially and economically [69]. Secure and transparent systems powered by blockchain link cotton farmers to better markets boost their negotiation power and guarantee fair payment for cotton produced using sustainable methods [70]. Also, through digital platforms, communities can exchange knowledge and decide together, which helps them cope better with environmental and economic shocks [71]. Taking all these changes together, they support poverty relief, better chances for earning a living, and the sustainability of rural communities which comes from making cotton production more practical, clear, and profitable for small farmers.

11. Case studies from leading cotton-growing countries

11.1 India: Digital Green and precision farming initiatives

Using digital videos and smartphones, Digital Green is helping cotton farmers in India get better extension services. With this method, people can share knowledge, practice sustainable farming, and achieve higher yields since it focuses on local environmental conditions and solves knowledge issues among smallholders [72]. With the help of IoT devices, drones, and GPS mapping in precision farming, less input is wasted and cotton yields and their fiber quality go up [73]. Because of Digital Green's approach and precision hardware, more sustainable and profitable cotton production can be achieved, increasing the productivity of Indian farms, helping the environment, and enhancing the lives of smallholders [74].

11.2 Pakistan: BCI and remote sensing in cotton

Major efforts to promote sustainable Pakistan cotton are led by the Better Cotton Initiative (BCI) in Punjab and Sindh. BCI gives training to smallholder farmers together with well-known organizations in water conservation, managing pests and maintaining healthy soil, to ease the impact on the environment and enhance both the quality and quantity of cotton produced (Better Cotton Initiative, n.d.). They help to improve the income of those working in the sector. At the same time, cotton areas are mapped and crop health can be observed by analyzing Sentinel-2 satellite pictures and using machine learning. It helps make decisions supported by data that can improve resource and input management [75]. Applying BCI's green standards and remote sensing technologies offers a firm plan to modernize Pakistan's cotton business and raise smallholder productivity, sustainability, and income (**Figure 2**).

11.3 USA: AI and robotics in commercial cotton farms

Farming cotton has been transformed by AI and robotics that handle many repetitive chores and use data to guide better choices. By examining information obtained from drones, sensors, and satellites, AI can track the health of crops, find pests at an early stage, and manage water and fertilizer better, raising yield and making resources used more effective [76]. Robotics like autonomous tractors, robotic weeders, and automated harvesters help big farms to work more efficiently and save labor costs [74]. The use of these technologies supports site-specific management of inputs in farming, which protects the environment and helps maintain evenly good cotton. By using AI in robotics, farmers are able to constantly observe and adjust to changes in weather and shortages of workers when producing cotton [44]. AI and robotics support sustainable growth and profitable results within the cotton industry.

Figure 2.
Initiatives for improved cotton production through integration of policies and digitization.

11.4 Africa: Low-cost technology solutions for small-scale growers

Small-scale cotton farmers in Africa have been helped by lower-cost technology which has increased their productivity, better use of resources and access to markets. The use of biochar in Zambia increases the fertility of the soil, raises cotton yields, and allows farmers to cut back on using chemical fertilizers, leading to sustainable farming [77]. Farmonaut in Zimbabwe provides smart farming services to smallholders at a lower cost, giving them advice on how to control insects and diseases [78]. Additionally, using digital literacy programs and mobile banking tools from Tech Herfrica, rural cotton growers are now able to connect with markets and get loans which improves their financial situation and allows them to afford enough food [79]. Through these improvements, little has been shown on how affordable and convenient technologies are transforming smallholder cotton farming, supporting development for society and protecting the environment in Africa.

12. Challenges and future directions

12.1 Barriers to digital adoption (technical, financial, educational)

Digital adoption in cotton farming faces several barriers, especially among smallholders. Technical challenges like poor internet connectivity and unreliable electricity in rural areas limit access to mobile apps, remote sensing, and IoT technologies [80]. Additionally, lack of localized software and language support reduces usability for diverse farmer groups. Financial constraints, including high upfront costs and limited access to credit, deter investment in digital devices and services [81]. Educational gaps, such as low digital literacy and limited awareness of digital agriculture benefits, further hinder adoption without adequate training and extension support [82]. Addressing these issues requires integrated approaches encompassing infrastructure improvements, affordable financing, localized content, and targeted capacity-building for cotton farmers.

12.2 Policy support and government initiatives

Policies and efforts by governments have a big impact on the sustainability, success, and performance of the global cotton market. Those at the forefront like the United States, India, China, Brazil, and Pakistan have implemented multiple approaches to help growers, protect the environment, and strengthen the cotton supply chain. Precision agriculture and biotechnology along with farm insurance policies that counter risk from weather and price changes are supported by American policies. Improving the quality of cotton fiber and helping the environment are aims reached through subsidies and research, as well as conservation schemes [62].

To help organic farming, India's government focuses on encouraging small farmers to join the market through the Cotton Development Programme. Besides, measures are taken to educate farmers, manage pests, and provide access to credits to help them produce more and improve their wealth [47]. In China, there is a strong drive to improve cotton farming using modern approaches and a lot of money is spent on research to improve the use of chemicals and water. The government helps develop certification schemes to meet the increasing domestic and foreign demand for ecologically grown cotton [65]. It reinforces production at scale using machinery and makes sure its producers care for the environment by joining the Better Cotton Initiative (BCI). Improving infrastructure, choosing suitable pest control, and fostering new markets are programs the government backs to improve cotton in the international market [83].

On an international scale, groups like the BCI and CmiA join forces with governments to support green cotton farming, better lives for farmers, and protect nature throughout the product value chain [84]. All these policies and actions enhance productivity, protect the environment, and lead to socioeconomic growth in regions where cotton is cultivated everywhere. Nevertheless, issues like global warming, changes in markets, and technology barriers keep appearing, so countries need to keep innovating their policies and cooperate on a global scale [76].

12.3 Emerging technologies (robotics, metaverse, digital twins)

Thanks to robotics, the metaverse, and digital twins, cotton farming's productivity, its level of sustainability and decision-making are expected to improve. With robotics, tasks such as planting, weeding, and harvesting can be done automatically, involving less human labor and helping increase efficiency with computer vision and AI [74]. The metaverse sets up virtual places for people to train, team up, and simulate activities, so cotton farmers and agricultural specialists can take virtual field tours, learn in special modules, and request consultations, helping expand the spirit of collaboration across the globe [41]. With the help of real-time sensor, drone and satellite data, digital twins copy cotton farms and predict changes in the crops, and expected diseases and how resources can be used well. By doing so, it is possible to examine how the management works under different conditions and improve both the sustainability and the harvest results [22]. All of these technologies combine to transform cotton farming, making it efficient, smart, and focused on the environment, while facing problems with labor, weather, and continuous changes [44].

13. Conclusions

Artificial Intelligence, IoT, blockchain, robotics, and big data analytics are likely to play a big role in the digital cotton sector of the future. The main objective is to grow more crops, harm the environment less, and benefit farmers by using cleaner farming practices. Through precision agriculture, sensors, drones, and satellites will monitor and adjust the use of water, nutrients, and pest control at each specific area. By using AI, farmers will be able to anticipate pest invasions, weather changes, and their harvest, so they can act ahead of time. New technologies such as blockchain and digital tags will secure the supply chain and allow farmers equal access to the market. With robotics and automation, issues related to planting, harvesting, and checking quality decrease. Because of virtual and augmented reality, information about farming can now be given to a wider audience. As time goes on, digital farming in the cotton industry will address climate problems, supply sufficient sustainable cotton worldwide, and aid farmers of all kinds.

Cotton cultivation has been transformed from labor intensive and heavy input driven into a data-driven, efficient, and sustainable system. Integration of precision technologies and equipment are making cotton growers more informed and critical decision makers to enhance productivity, profits, and environmental stewardship. However, the success depends upon the technology access, technology adoption, farmers training, and supportive policies.

DOI: http://dx.doi.org/10.5772/intechopen.1011733

Author details

Ghulam Sarwar*, Muazzama Batool and Muhammad Tauseef
Cotton Research Institute, Multan, Pakistan

*Address all correspondence to: sarwar87@yahoo.com

References

[1] Ahmad S, Hasanuzzaman M, editors. Cotton Production and Uses: Agronomy, Crop Protection, and Postharvest Technologies [Internet]. Singapore: Springer Singapore; 2020. Available from: http://link.springer.com/10.1007/978-981-15-1472-2 [Accessed: February 20, 2024]

[2] Jabran K, Ul-Allah S, Chauhan BS, Bakhsh A. An introduction to global production trends and uses, history and evolution, and genetic and biotechnological improvements in cotton. Cotton Production. 2019:1-22

[3] Ansell MP, Mwaikambo LY. The structure of cotton and other plant fibres. In: Handbook of Textile Fibre Structure. Elsevier; 2009. pp. 62-94

[4] Wendel JF, Grover CE. Taxonomy and evolution of the cotton genus, gossypium. Cotton. 2015;**57**:25-44

[5] Wegier A, Alavez V, Piñero D. Cotton: Traditional and modern uses. In: Ethnobotany of Mexico: Interactions of People and Plants in Mesoamerica. Springer; 2016. pp. 439-456

[6] Fang E, Lee J, Yang Z. The timing of codevelopment alliances in new product development processes: Returns for upstream and downstream partners. Journal of Marketing. 2015;**79**(1):64-82

[7] Shuli F, Jarwar AH, Wang X, Wang L, Ma Q. Overview of the cotton in Pakistan and its future prospects. Pakistan Journal of Agricultural Research. 2018;**31**(4):396

[8] Pretty J, Benton TG, Bharucha ZP, Dicks LV, Flora CB, Godfray HCJ, et al. Global assessment of agricultural system redesign for sustainable intensification. Nature Sustainability. 2018;**1**(8):441-446

[9] Devlet A. Modern agriculture and challenges. Frontiers in Life Sciences and Related Technologies. 2021;**2**(1):21-29

[10] Esmaeili L, Hashemi GSA. A systematic review on social commerce. Journal of Strategic Marketing. 2019;**27**(4):317-355

[11] Abdelraheem A, Esmaeili N, O'Connell M, Zhang J. Progress and perspective on drought and salt stress tolerance in cotton. Industrial Crops and Products. 2019;**130**:118-129

[12] Nechaev VP, Bechtel A, Dai S, Chekryzhov IY, Pavlyutkin BI, Vysotskiy SV, et al. Bio-geochemical evolution and critical element mineralization in the cretaceous-Cenozoic coals from the southern Far East Russia and northeastern China. Applied Geochemistry. 2020;**117**:104602

[13] Nechaev A, Tyapkina M, Kireeva E. Development of innovative technologies in the agricultural industry. EDP Sciences. 2023:01043

[14] Kamble B, Kilic A, Hubbard K. Estimating crop coefficients using remote sensing-based vegetation index. Remote Sensing. 2013;**5**(4):1588-1602

[15] Khobragade P, Dhankar PK, Titarmare A, Dhone M, Thakur S, Saraf P. Quantum-Enhanced AI Robotics for Sustainable Agriculture: Pioneering Autonomous Systems in Precision Farming. IEEE; 2024. pp. 1-7

[16] Zhumaxanova K, Yessymkhanova Z, Yessenzhigitova R, Kaydarova A. The current state of agriculture digitalization: Problems and ways of solution. Central Asian Economic Review. 2019;**5**(128):145

[17] Basso B, Antle J. Digital agriculture to design sustainable agricultural systems. Nature Sustainability. 2020;**3**(4):254-256

[18] Himesh S, Rao EP, Gouda K, Ramesh K, Rakesh V, Mohapatra G, et al. Digital revolution and big data: A new revolution in agriculture. CABI Reviews. 2018;**2018**:1-7

[19] Stočes M, Vaněk J, Masner J, Pavlík J. Internet of things (iot) in agriculture-selected aspects. Agris on-line Papers in Economics and Informatics. 2016;**8**(1):83-88

[20] Kim YL. Data-driven approach to characterize urban vitality: How spatiotemporal context dynamically defines Seoul's nighttime. International Journal of Geographical Information Science. 2020;**34**(6):1235-1256

[21] Li XH, Cao CC, Shi Y, Bai W, Gao H, Qiu L, et al. A survey of data-driven and knowledge-aware explainable AI. IEEE Transactions on Knowledge and Data Engineering. 2020;**34**(1):29-49

[22] Zhang L, Dabipi IK, Brown WL Jr. Internet of things applications for agriculture. Internet of Things A to Z: Technologies and Applications. 2018:507-528

[23] Ullo SL, Sinha GR. Advances in IoT and smart sensors for remote sensing and agriculture applications. Remote Sensing. 2021;**13**(13):2585

[24] Eli-Chukwu NC. Applications of artificial intelligence in agriculture: A review. Engineering, Technology and Applied Science Research. 2019;**9**(4)

[25] Bill R, Nash E, Grenzdörffer G. GIS in Agriculture. In: Springer Handbook of Geographic Information. Springer; 2012. pp. 461-476

[26] Qiang CZ, Kuek SC, Dymond A, Esselaar S, Unit I. Mobile applications for agriculture and rural development. Washington DC: World Bank; 2012

[27] Costopoulou C, Ntaliani M, Karetsos S. Studying mobile apps for agriculture. IOSR Journal of Mobile Computing & Application. 2016;**3**(6):44-49

[28] Dibbern T, Romani L, Massruhá S. Main drivers and barriers to the adoption of digital agriculture technologies. Smart Agricultural Technology. 2024;**8**:100459

[29] Buiak L. Modern trends and main theoretical approaches to the digital transformation of agribusiness. Journal of Strategic Economic Research. 2023;**6**(24):50-62

[30] Kumar P, Joshi P, Birthal PS. Demand projections for foodgrains in India. Agricultural Economics Research Review. 2009;**22**(2):237-244

[31] Gürsoy S. A review of the factors affecting adoption of precision agriculture applications in cotton production. Best Crop Management and Processing Practices for Sustainable Cotton Production. 2024

[32] Huang Y, Reddy KN, Thomson SJ, Yao H. Assessment of soybean injury from glyphosate using airborne multispectral remote sensing. Pest Management Science. 2015;**71**(4):545-552

[33] Xun L, Zhang J, Yao F, Cao D. Improved identification of cotton cultivated areas by applying instance-based transfer learning on the time series of MODIS NDVI. Catena. 2022;**213**:106130

[34] Elms MK, Green CJ, Johnson PN. Variability of cotton yield and quality. Communications in Soil Science and Plant Analysis. 2001;**32**(3-4):351-368

[35] Nair S, Wang C, Segarra E, Johnson J, Rejesus R. Variable rate technology and cotton yield response in Texas. Journal of Agricultural Science and Technology B. 2012;**2**(9B):1034

[36] Tantalaki N, Souravlas S, Roumeliotis M. Data-driven decision making in precision agriculture: The rise of big data in agricultural systems. Journal of Agricultural & Food Information. 2019;**20**(4):344-380

[37] Narra N, Nevavuori P, Linna P, Lipping T. A data driven approach to decision support in farming. In: Information Modelling and Knowledge Bases XXXI. IOS Press; 2020. pp. 175-185

[38] Hardin RG IV, Barnes EM, Delhom CD, Wanjura JD, Ward JK. Internet of things: Cotton harvesting and processing. Computers and Electronics in Agriculture. 2022;**202**:107294

[39] Zhao M, Liu W, Saif ANM, Wang B, Rupa RA, Islam KA, et al. Blockchain in online learning: A systematic review and bibliographic visualization. Sustainability. 2023;**15**(2):1470

[40] Godhani K, Patel A, Thakkar D, Shah M. Application of remote sensing in smart agriculture using artificial intelligence. In: Agricultural Biotechnology. CRC Press; 2022. pp. 357-370

[41] Li J, Zhao M, Qin Z, Yuan R, Huang A, Li M. Detection and recognition of multiple QR codes based on YOLO_CBAM algorithm. International Journal of Bio-Inspired Computation. 2024;**23**(3):179-188

[42] Alkhayyat A, Kumar R, Singh S, Singh R, Kumar Y, Sharma U. An Online QR Code Scanner for Real-Time User Feedback and Ratings Collection with Local Web Server. IEEE; 2024. pp. 733-736

[43] Shah P, Patel CR. Prevention is better than cure: An application of big data and geospatial technology in mitigating pandemic. Transactions of Indian National Academy of Engineering. 2020;**5**(2):187-192

[44] Kamilaris A, Kartakoullis A, Prenafeta-Boldú FX. A review on the practice of big data analysis in agriculture. Computers and Electronics in Agriculture. 2017;**143**:23-37

[45] Zhang X, Zhou M. Factors affecting university students' intention to use mobile apps for intercultural competence development: A mixed-method approach. Education and Information Technologies. 2024;**29**(6):6991-7026

[46] Klerkx L, Jakku E, Labarthe P. A review of social science on digital agriculture, smart farming and agriculture 4.0: New contributions and a future research agenda. NJAS - Wageningen Journal of Life Sciences. 2019;**90**:100315

[47] Kumar L, Ahlawat P, Rajput P, Navsare R, Singh PK. Internet of things (IOT) for smart precision farming and agricultural systems productivity: A review. International Journal of Engineering Applied Science and Technology. 2021;**5**:141-146

[48] Owusu-Peprah NT. World Development Report 2022: Finance for an Equitable Recovery: By World Bank. Washington, DC: World Bank Publications; 2022. p. 267 $0.00 (E-book), ISBN 9781464817311. Vol. 38. Taylor and Francis; 2024

[49] Aker JC, Mbiti IM. Mobile phones and economic development in Africa. The Journal of Economic Perspectives. 2010;**24**(3):207-232

[50] Mittal S. The role of mobile phones in empowering women in

agriculture. In: Routledge Handbook of Gender and Agriculture. Routledge; 2020. pp. 160-169

[51] Cole SA, Fernando AN. 'Mobile'izing agricultural advice technology adoption diffusion and sustainability. The Economic Journal. 2021;**131**(633):192-219

[52] Tadesse G, Bahiigwa G. Mobile phones and farmers' marketing decisions in Ethiopia. World Development. 2015;**68**:296-307

[53] Casaburi L, Kremer M, Mullainathan S, Ramrattan R. Harnessing ICT to Increase Agricultural Production: Evidence from Kenya. Harvard University; 2014. p. 2

[54] Okeleke K. GSMA National Dialogues. 2021

[55] Nakasone E, Torero M. A text message away: ICTs as a tool to improve food security. Agricultural Economics. 2016;**47**(S1):49-59

[56] De Siqueira AC. The 2020 World Bank education strategy: Nothing new, or the same old gospel. In: The World Bank and Education. Brill; 2012. pp. 67-81

[57] Swanson BE, Rajalahti R. Strengthening agricultural extension and advisory systems: Procedures for assessing, transforming, and evaluating extension systems. Agriculture and Rural Development Discussion Paper. 2010;**45**:1-187

[58] Aker JC, Boumnijel R, McClelland A, Tierney N. Zap it to me: The short-term impacts of a mobile cash transfer program. Center for Global Development Working Paper. 2011:268

[59] Fawole OP, Olajide BR. Awareness and use of information communication technologies by farmers in Oyo state, Nigeria. Journal of Agricultural & Food Information. 2012;**13**(4):326-337

[60] Birner R, Daum T, Pray C. Who drives the digital revolution in agriculture? A review of supply-side trends, players and challenges. Applied Economic Perspectives and Policy. 2021;**43**(4):1260-1285

[61] Mekouar MA. Food and agriculture Organization of the United Nations (FAO). Yearbook of International Environmental Law. 2023;**34**(1):yvae031

[62] Zhao N, Zhou L, Huang T, Taha MF, He Y, Qiu Z. Development of an automatic pest monitoring system using a deep learning model of DPeNet. Measurement. 2022;**203**:111970

[63] Singh A, Mehrotra R, Rajput VD, Dmitriev P, Singh AK, Kumar P, et al. Geoinformatics, artificial intelligence, sensor technology, big data: Emerging modern tools for sustainable agriculture. Sustainable Agriculture Systems and Technologies. 2022:295-313

[64] Liu Z, Li J. Application of unmanned aerial vehicles in precision agriculture. Agriculture. 2023;**13**(7):1375

[65] Wang J, Wang Y, Li G, Qi Z. Integration of remote sensing and machine learning for precision agriculture: A comprehensive perspective on applications. Agronomy. 2024;**14**(9):1975

[66] Garcia DC, Cangirana LS, de Queiroz RG, Gimenes RMT. Determining factors in the adoption of precision agriculture technologies by grain producers. Informe Gepec. 2024;**28**(1):384-399

[67] Patel PK, Kumar H, Patel J. Determinant Factors for Social Media Utilisation Behaviour of Farmers for

Sustainable Agriculture Development. 2023

[68] Aker JC. Cash or coupons? Testing the impacts of cash versus vouchers in the Democratic Republic of Congo. Center for Global Development Working Paper. 2013:320

[69] Ciaian P. Use of Blockchain Applications in the Agri-Food Sector: State of Play. 2020

[70] Voora V, Bermudez S, Farrell JJ, Larrea C, Luna E. Cotton prices and sustainability. International Institute for Sustainable Development. 2023:1-37

[71] Wossen T, Abdoulaye T, Alene A, Haile MG, Feleke S, Olanrewaju A, et al. Impacts of extension access and cooperative membership on technology adoption and household welfare. Journal of Rural Studies. 2017;**54**:223-233

[72] Raj M, Seamans R. Primer on artificial intelligence and robotics. Journal of Organization Design. 2019;**8**(1):11

[73] Singh R, Patel N, Danodia A. Mapping of sugarcane crop types from multi-date IRS-Resourcesat satellite data by various classification methods and field-level GPS survey. Remote Sensing Applications: Society and Environment. 2020;**19**:100340

[74] Shamshiri R, Weltzien C, Hameed IA, Yule JI, Grift ET, et al. Research and Development in Agricultural Robotics: A Perspective of Digital Farming. 2018

[75] Goheer M, Hassan S, Sheikh A, Malik Y, Uzair M, Satti T. Assessing smog trends and sources of air pollutants across northeastern districts of Punjab, Pakistan using geospatial techniques. International Journal of Environmental Science and Technology. 2025;**22**(5):3657-3674

[76] Liakos KG, Busato P, Moshou D, Pearson S, Bochtis D. Machine learning in agriculture: A review. Sensors. 2018;**18**(8):2674

[77] Mwansa-Kambafwile JR, Jewett S, Chasela C, Ismail N, Menezes C. Initial loss to follow up of tuberculosis patients in South Africa: Perspectives of program managers. BMC Public Health. 2020;**20**(1):622

[78] Richardson RB, Olabisi LS, Sakana N, Waldman K, Grabowski P. The Impact of Sustainable Intensification on Landscapes and Livelihoods (SILL) in Zambia. 2015

[79] Adeyemi J. Contemporary Art from Nigeria in the Global Markets: Trending in the Margins. Springer; 2022

[80] Ojo J, Ajiboye J, Ajiboye M, Ajiboye D, Ohize H, Isa A. Design and implementation of real-time internet of things (Iot) enhanced irrigation system. El-Amin University Minna. 2024;**1**(1):31

[81] Kassie M, Tesfaye K. Malting barley grain quality and yield response to nitrogen fertilization in the Arsi highlands of Ethiopia. Journal of Crop Science and Biotechnology. 2019;**22**(3):225-234

[82] Gebremedhin B, Hoekstra D, Jemaneh S. Heading towards Commercialization? The Case of Live Animal Marketing in Ethiopia. 2007

[83] Santos E, Fernandes CI, Ferreira JJ. The role of political and economic institutions in informal entrepreneurship. World Journal of Entrepreneurship, Management and Sustainable Development. 2019;**15**(4):366-383

[84] Pallavi S. A Study on the Impact of Better Cotton Initiative (BCI) Programme on Cotton Growers in Karimnagar District of Telangana State. MSc(Ag) Thesis. 2016

Section 3

Textile, Cotton Products and Use

Chapter 5

Composites Derived from Cotton Wastes: Study of Microscopic and Macroscopic Properties of Contemporary Materials

Vignesh Natarajan

Abstract

Cotton waste, known for its distinguishable mechanical properties and high crystallinity, is an extensively investigated additive in the manufacture of polymer composites. The blending of cotton waste modulates its physiochemical properties, such as providing recalcitrance to soft materials and extending its flexibility to hard materials. However, the preprocessing of cotton waste into a desirable consistency (particle size) is essential for enhancing the intermolecular adhesive interactions and elevating the durability of the composites. The chapter covers key aspects of the structural reinforcements and functional modifications of cotton-blended polymers for microscopic and macroscopic applications in biomedical and civil engineering sectors, respectively. The blending of various parts of waste cotton on the bulk properties and the hydrophilicity of composites has been put forth with supporting information from proof-of-concepts and detailed literature reports. The specific addition of waste cotton to polymer composites has significantly reduced the thermal conductivity. In contrast, pronounced elevation in the sound absorption and moisture retention properties was reported for waste cotton reinforced fiberboards. Hence, the chapter intends to cover the most recent state-of-the-art literature describing the versatile effects of cotton composition and cotton preprocessing on the strength and resilience of contemporary materials of common usage.

Keywords: cotton waste, preprocessing methods, soft materials, hard materials, physiochemical properties, composites

1. Introduction

Cellulose, a long linear chain consisting of hydroxyl groups, is the structural framework providing osmotic tolerance, rigidity, and cell attachment sites in plants. Due to the abundance of hydroxyl groups and the existence of numerous hydrogen bonding networks, cellulose serves as the foundation for the organization of complex plant tissue architecture [1]. Moreover, the non-covalent adsorption of lignin, hemicellulose, and pectin with cellulose differs between parenchyma, chlorenchyma, and sclerenchyma. As a result, the permeability of the cell wall to chemicals, water, and proteins is widely different in the tissues underlying the root, stem, and leaf of the plant [2]. Prevalently, the cellulose content in the plant tissue ranges between 20 and 40% in agricultural residues. However, the cotton plant is exceptional owing to the diversity in the cellulose content. For instance, cotton stalks and hulls consist of less than 60% cellulose. On the other hand, cotton bolls have a cellulose content of 90% on average. Notably, the crystallinity of cellulose is a distinguishing factor between cotton stalks and bolls [3]. Cotton is one of the few plant varieties that provide wholesome utilization of all the plant parts (stem, leaf, bolls, seeds) for multiple commercial applications. The most common application is the refinement and spinning of cellulose into elongated threads of unprecedented tensile strength and tenacity index. The crystallinity of cellulose in cotton shares an immense contribution in elevating the mechanical toughness of textile fibers [4].

Although cotton cellulose is widely used in textiles, food packaging, fiber production, and drug delivery systems, it lacks the functional groups necessary for supporting the regeneration of animal tissues [5]. Moreover, the lack of antimicrobial activity stands out as a significant concern in the development of cellulose-based wound dressing materials. Cellulose in its native form shows poor dispersion in other polymer matrixes during the fabrication of multifunctional composite materials [6, 7]. Moreover, the harvested cotton is a complex mixture of bolls, stems, motes, and seeds, and the waste cotton discharged after cellulose refinement is heterogeneous in composition [3, 4]. Thus, alkali bleaching and sulfite treatments enhance the homogeneity of the discharged waste and increase the cellulose fiber content. The initial refining process includes mechanical sieving to separate particles of higher density and weight (seeds, hulls, and motes). Further, the use of alkali dissolves the phenolic compounds and amorphous components. Nevertheless, the alkali-treated cellulose requires further preprocessing before the final preparation of cotton cellulose-based composite materials [8]. Importantly, the breakdown of cellulose's degree of polymerization is indispensable to extend the application of cotton fibers in the preparation of soft and durable composite materials. Mechanical disintegration is simple with respect to the recovery of cellulose fibrils of narrow polydispersity index. However, repeated defibrillation of microfibers to nanofibrils is energy demanding. The cellulose nanofibrils possess microscopic physiochemical properties similar to those of the fiber yarn in textile fabric. Notably, the fibrils are semi-crystalline with the occurrence of ordered and disordered regions in the cellulose I structure [9, 10].

Likewise, alkali pretreatment generates semi-crystalline cellulose through hydrolysis of ester linkages in the lignin-carbohydrate complex. On the contrary, hydrolysis using concentrated acid generates highly crystalline cellulose. Moreover, the cellulose crystals undergo chemical modification at the C6-hydroxyl groups. Though the crystallinity of the cellulose is increased considerably, the introduction of polar or charged functional groups in the modified cellulose elevates the dispersion and distribution coefficients in copolymer matrices [9, 11]. On the other hand, regeneration of cellulose using ionic liquids, sodium hydroxide, urea, and molten inorganic salt introduces the self-annealing and thermostable properties to develop an all-cellulose composite [12, 13]. Besides, the regenerated cellulose preparation reverses the intermolecular hydrogen bonding network in cellulose microfibrils. More prevalently, the parallel arrangement of cellulose microfibrils is disrupted by ionic liquids and molten inorganic salts. Further, the displacement of ionic liquid or cellulose solvents rearranges cellulose microfibrils in antiparallel conformation [12]. As an alternative, carbonization of cotton cellulose in an oxygen-deficient environment alters the elemental properties due to the decomposition of hydrogen, oxygen, and other volatile matter. For instance, thermochemical decomposition of cellulose by pyrolysis removes the chemical groups, causing the intricate packing of cellulose into standalone fibers [14]. Moreover, impregnation of the pyrolyzed biomass in textile fibers, dental implants, and wound dressing material has resulted in the augmentation of moisture absorption, mitigation of tooth decay, and faster wound healing, respectively [15–17]. Hence, the preprocessing of cellulose widens the scope of applicability in biomedical, civil engineering, and other miscellaneous areas of research and development [18]. The present chapter provides detailed evidence of preprocessing methods' influences on the cotton cellulose-derived composite properties.

2. Influences of preprocessing methods on the physiochemical properties of cotton waste

Cotton waste, though primarily composed of cellulose, has been observed to differ in the biomass composition depending on the harvesting season in the fields and refining methods in the textile mills [3]. The heterogeneous composition of cotton waste arises due to the occurrence of cotton motes, stalks, and seeds in the unrefined secondary discharge fraction. Moreover, the discharge fractions arising from the cotton yarn production in the ginning and spinning operations are associated with waste management challenges. Thus, the valorization of secondary cotton waste into resourceful products alleviates the carbon footprint [4, 19]. Typically, the cotton waste is subjected to preprocessing by size reduction and chemical treatments to enhance the dispersion in several polymer composites (**Figure 1**).

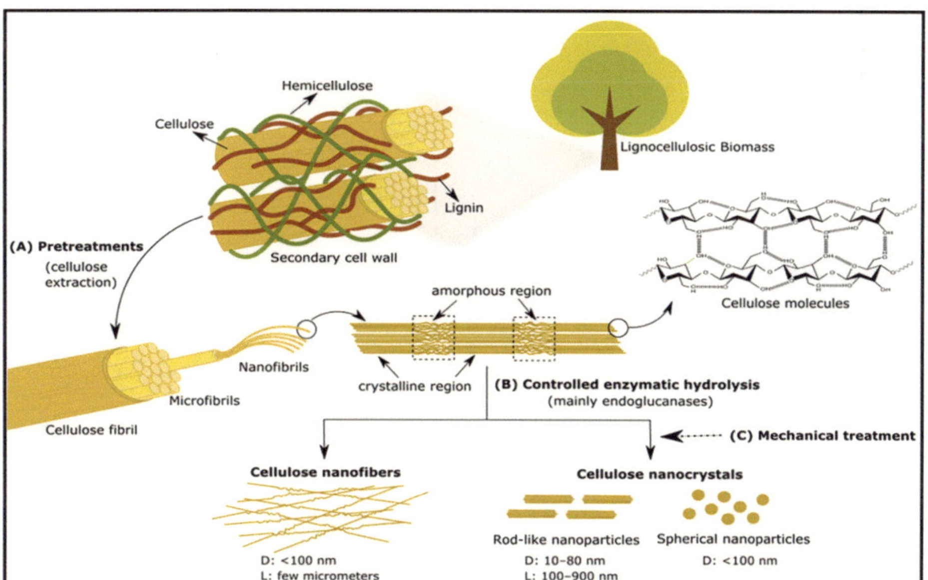

Figure 1.
Preprocessing of cellulose into nanofibers and nanocrystals through mechanical and chemical treatments (reproduced with copyright permission from [20], MDPI open access journal).

Mechanical fractionation includes the splitting of cotton fibers into cellulose nanofibrils through high-pressure homogenization and microfluidization. Mechanical fibrillation has been widely achieved by ball milling, extrusion, and ultrasonication [9, 10]. The benefits of mechanical cellulose size reduction include the extraction of unmodified cellulose in its crystalline form for its application in tissue scaffold development. Further, cross-linking cellulose nanofibrils using divalent metal ions leads to the formation of hydrogels. Retention of the fiber length-to-diameter ratio in the cellulose nanofibrils facilitates the fabrication of standalone hydrogels for wound dressing applications. Mechanical processing of cellulose into nanofibrils is efficient, provided the degree of polymerization of the cellulose is reduced [5–7, 9]. The physical pretreatment of sugarcane bagasse by steam explosion increased the yield of 3–7 nm cellulose fibrils in the sequential processing step of high-pressure microfluidization [21]. In contrast, acid hydrolysis of cotton waste into nanocrystals was essential for further degradation of crystalline cellulose into nanofibrils by the high-pressure homogenization [22]. The critical optimization of acid hydrolysis and homogenization resulted in a 72% yield of cellulose nanocrystals derived from cotton linter, which has twice the water-holding capacity. Further, inclusion of an ultrasonication step gave rise to the formation of well-dispersed nanoparticles with dimensions of 133 nm × 10 nm [23]. Researchers demonstrated that combining electron beam irradiation can effectively break down cellulose in waste cotton fabric into cellulose nanocrystals, serving as a replacement for hydrolysis with concentrated sulfuric acid. Furthermore, the novel route of cellulose nanocrystal preparation introduced emulsifying properties and opened up new applications in the formulation of Pickering emulsions [24].

Chemical treatments of cotton wastes, on the other hand, likely generate regenerated cellulose due to the reconstruction and redistribution of intermolecular hydrogen bonds (**Figure 2**). Commonly, the chemical treatments elevating the solubilization

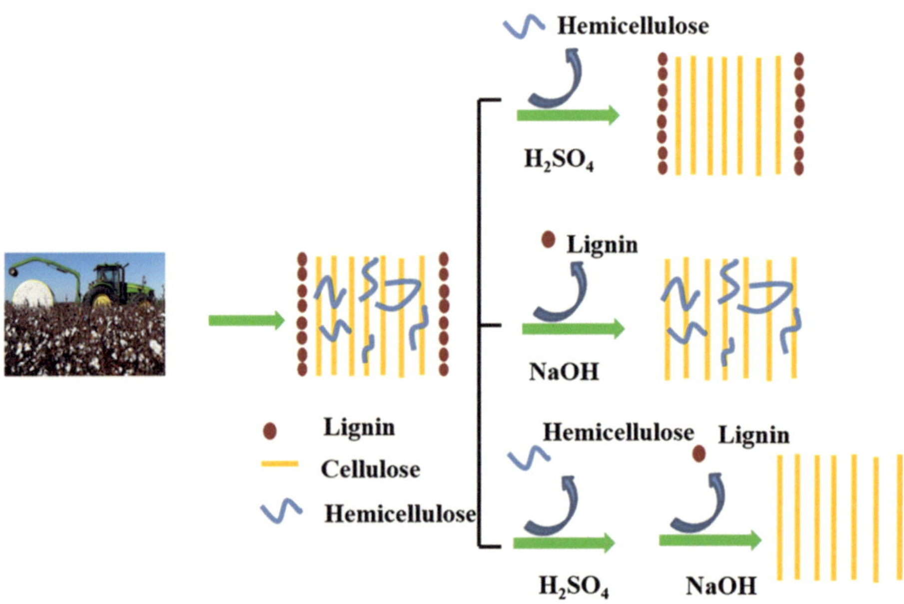

Figure 2.
The isolation of cellulose from cotton waste residues by alkali and acid treatments (reproduced with copyright permission from [25], Springer Open access).

of cotton waste include ionic liquid solvation, molten inorganic salt dissolution, and alkali bleaching. However, the crystallinity of the cotton waste is disrupted, and the hydrophilicity of the regenerated cellulose becomes prominent. Nevertheless, ionic liquids facilitate the selective extraction of crystalline cellulose from cotton gin trash and contribute to the enhancement of the Young's modulus of the regenerated cellulose fibers [12, 13]. For instance, the dissolution process using 1-allyl-3-methylimidazolium chloride significantly altered the rheology, tensile properties, and thermal stability of the regenerated cellulose fiber. Compared to wood pulp, regenerated cellulose from cotton waste showed a larger tensile strength, an elevated degree of polymerization, and improved thermal stability. Though cotton waste was deemed to be more resilient, the total crystallinity index of regenerated cellulose was similar to that of wood pulp. In contrast, the modulus of elasticity of regenerated cellulose from cotton waste was 47.52% higher than that of wood pulp [26]. N-methyl morpholine N-oxide (NMMO) has been extensively used as a non-toxic solvent for aiding in the dissolution of cellulose in wood pulp. The tenacity of the regenerated cellulose arising from the NMMO-solubilized wood pulp is comparable to that of pure cotton's cellulose. On the contrary, the blending of cotton waste and wood pulp in the ratio of 1:4 was reported to result in a composite fiber with a cumulative increase in the tenacity index and wet strength recovery [27]. Lithium chloride salt, as a result of complementary interaction with N, N-dimethylacetamide, creates a hydration shell and causes the solvation of crystalline cellulose. The release of protons during the electrostatic interaction facilitates the breakdown of glycosidic linkage in cellulose microfibers. However, partial solvation of medical-grade cotton in dissolved microcrystalline cellulose matrix solution improved the tensile strength and crystallinity of the self-reinforced composite film. The composite film's properties emphasized its significance in the fabrication of fracture plaster [28].

Thermochemical treatments introduce self-reliant and novel properties in the pretreated cotton waste. For instance, pyrolysis of cotton waste produced biochar due to the thermal decomposition in an oxygen-deficient environment [14]. The yield of biochar was reported to be significant, as at least two-thirds of cotton waste's dry weight consisted of cellulose (**Figure 3**). The confined heating range between 25 and 200°C provided the route for the manufacture of fiber-intertwined biochar with the retention of biochar's intermolecular adsorption property and cotton fiber's tensile strength. The heating between 150 and 200°C was shown to be critical in the specific decomposition of hemicellulose. Moreover, the non-bonding interactions in amorphous cellulose, such as Van der Waals forces, Coulomb forces, and hydrogen bonds, preserved the tensile properties of the fiber up to 200°C [30]. In a wider heating range from 300–500°C, the fixed carbon content, porosity, and calorific value of waste cotton showed significant rises at specific temperatures of 500 and 400°C, respectively. Though the yield of biochar was observed to reduce from 50.5 to 26.7% in the above heating range, the fuel value of the biochar and the graphite formation were elevated. Considerable reduction in the weight of biochar represented the presence of high volatile content (cellulose and hemicellulose) in the waste cotton [14]. Owing to the rapid decline in the solid weight of biochar at temperatures above 500°C, microporous activated carbon was obtained at 450°C for the acid-impregnated cotton waste. The initial soaking in 30% phosphoric acid elevated the micropore volume, and higher impregnation ratios facilitated the formation of mesopores at a lower heating temperature of 450°C in the pyrolysis process [31]. The printing of biochar as an additive on textile fabrics has been shown to increase moisture transfer and enlarge water vapor permeability, but alleviate oxygen transfer

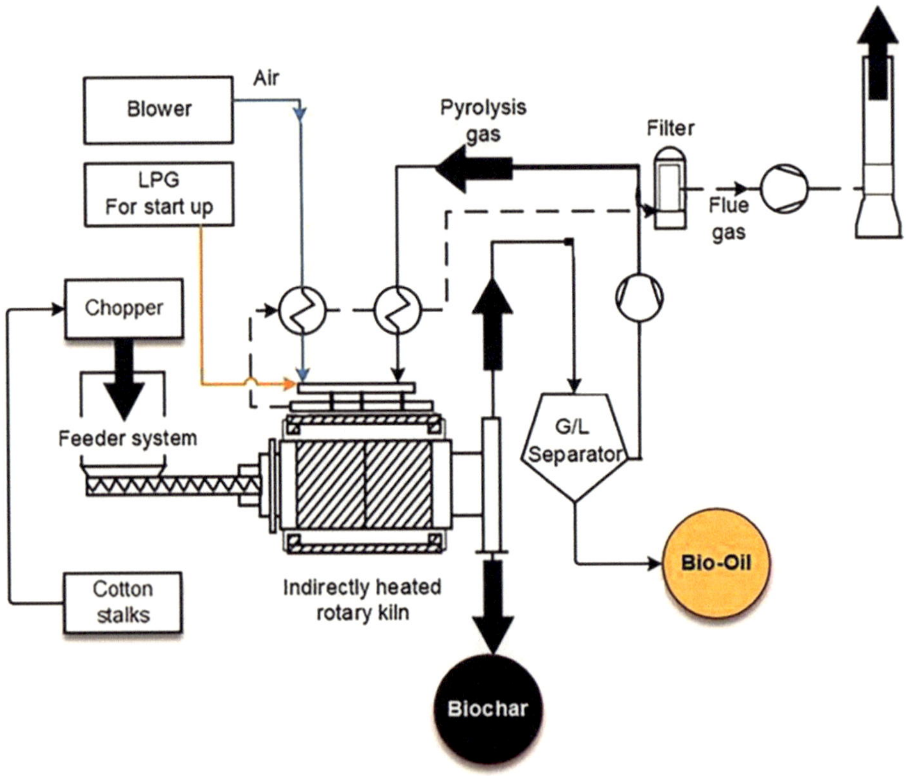

Figure 3.
Thermochemical decomposition of cotton waste yields biochar rich in amorphous cellulose fraction (reproduced with copyright permission from [29], IntechOpen book chapter).

and gas exchange [32]. Likewise, the impregnation of biochar has been reported to mitigate electromagnetic interference and augment the dielectric property in the silk fiber-reinforced vinyl ester composite film [33].

3. Composites in biomedical engineering: Blending of cotton waste on microscopic properties

Cotton waste, arising from chemical treatments and mechanical homogenization, exhibits strong interfacial adhesion properties [9]. The breaking of intermolecular hydrogen bonding in cotton cellulose is essential for widening the dispersion of nanofibrils in the copolymer solutions (**Figure 4**). Further, the blending of cellulose with a matrix has been reported to modify the elasticity and impart tenacity to the hydrogels and liquid-filled polymers. Specifically, the incorporation of alkali-treated cellulose from cotton waste bridges the intermolecular distance between the linear chains and increases the hydrogen bonding interactions to provide mechanical stiffness [22–24].

In the initial preparation of antibacterial wound dressing material, alkali-treated cellulose from waste cotton was reacted with 34% carbon disulfide to synthesize cellulose xanthate. Further, the blending of chitosan and its interfacial adhesion with cellulose xanthate facilitated the viscose fiber functionalization and the

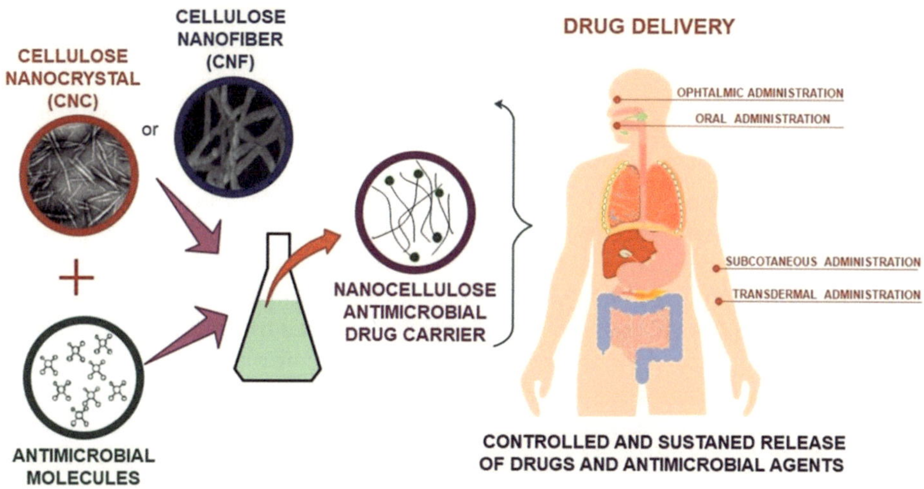

Figure 4.
Significance of nanocellulose derived from cotton waste on the development of drug delivery systems (reproduced with copyright permission from [34], MDPI open access journal).

emergence of bacterial resistance. An average degree of polymerization of the viscose fiber (782) correlated with the solution viscosity of 880 to 1495 centipoise. Owing to the high crystallinity and degree of polymerization of cotton cellulose, the treatments using multiple chemicals still favored the wet-spinning of chitosan-functionalized viscose fiber into a durable and extendable sanitary pad [35]. Rice husk, a prominent biomass with high cellulose content, has been widely investigated for ethanol production. Alternatively, the cellulose dissolution by sodium hydroxide has been applied as a common step for reinforcing the industrial cotton waste with rice husk. A full-fledged wound dressing material was further formulated through the additional incorporation of silver nitrate nanoparticles. However, the preliminary extraction of cellulose from rice husks by alkali-assisted bleaching was indispensable. The reinforcement of industrial cotton waste with rice husk significantly increased the fluid absorbency (1282%), tensile strength (28.6 N), and elongation at break (41%) of the wound dressing material [36].

The addition of cotton waste has been reported to alter the air permeability and moisture absorption in biodegradable composite hydrogels. The prominent increment in gas permeability (108 mm s^{-1}) was observed during the incorporation of turmeric into the chitosan/alginate polyelectrolyte hydrogel. The observations of 420% moisture absorption, 14% stretchability, and 32 N tensile strength were attributed to the dense cross-linking network between polyelectrolyte hydrogel and industrial cotton non-woven fabric. Moreover, calcium chloride supplementation in the hydrogel caused aggregation of the polymer matrix through semi-interpenetrating network formation [37]. The high molecular weight of cellulose in cotton waste facilitated the production of a higher yield of cellulose nanocrystals *via* acid hydrolysis. In addition, dispersion of cellulose nanocrystal in the polycaprolactone augmented the biodegradability of the electrospun nanofibrous mat of 233 nm diameter. Importantly, the incorporation of cellulose nanocrystals retarded the release of tetracycline from the nanofibrous mat and contributed to increased drug sequestration [38].

The fluorescent sensors have been widely used in the *in vivo* detection of diseases and antigenic biomarkers at the subcellular level. However, several aromatic fluorescent molecules are immunogenic, and these have been replaced by the carbon dots. The latter has been derived through thermochemical decomposition of plant biomass. However, cotton waste is the preferred source for carbon dot preparation, owing to the high cellulose content [14]. The distinguishing feature of carbon dots is the display of photoluminescence in the visible and near-infrared regions. Although the critical observation indicated that the carbon dots produced from acetylated cellulose in waste cotton are compatible with human dermal fibroblast cells, they also exhibited irradiation-dependent antibacterial action against Gram-negative *Escherichia coli*. Furthermore, the impregnation of carbon dots on polyvinyl alcohol introduced the phenomenon of photocatalytic effect quenching to track the antibacterial action [39]. In another study, co-precipitation of iron oxide nanoparticles and cellulose nanofibril hydrogel preparation was carried out at sub-zero temperature using a sodium hydroxide-urea aqueous two-phase system. The encapsulation of iron oxide was further demonstrated through chemical cross-linking using citric acid and physical cross-linking by the freeze-thaw process. The major finding emphasized the magnetic field-dependent release of iron oxide due to the presence of magnetic hypothermia-induced deformations in the cross-linking network [40]. Iron oxide nanoparticles derived from the spent tea waste were integrated with cotton cellulose to improve the binding affinity of the antibiotic metronidazole. The increased drug sequestration in the nanocomposite was attributed to its larger swelling capacity than that of pure cellulose. The encapsulation of iron oxide in the nanocellulose composite caused a decline in its magnetic property, which was responsible for the sustained drug release [41].

Tissue engineering for organ transplantation includes the development of scaffolds to support the regeneration of the tissue's microenvironment [5]. Chiefly, cellulose from cotton waste has been incorporated to increase the water retention and aid in the exchange of gases and nutrients across the scaffold's porous matrix (**Figure 5**). Cellulose microfiber from the surgical cotton was blended with gelatin, egg white, and nanohydroxyapatite through complex chemical cross-linking using 1-ethyl-3-(3-dimethylaminopropyl) carbodiimide and N-hydroxysuccinimide. However, gelatin served as a functional polymer for supporting adhesion in osteoblast cells. Further, structural analysis by field emission scanning electron microscopy and atomic force microscopy confirmed the availability of numerous cell attachment sites on the surface of egg white protein. Nevertheless, the significance of cellulose microfiber addition corresponded to the increase in the compressive modulus and average pore size of the composite scaffold [43]. Cotton cellulose treated through acid hydrolysis and probe sonication served as an adhesive matrix for the bone tissue engineering when supplemented with the bone-conditioned medium. The breakdown of finer fibril structure by probe sonication was observed to improve the cellulose functionalization with bone-conditioned medium [44]. Cellulose nanocrystals (CNC) derived from industrial cotton waste acted as mechanical reinforcement in the development of polycaprolactone/sodium alginate mat. Moreover, the cross-linking by both CNC and sodium alginate was shown to promote the adhesion of NIH/3 T3 cells. The high-water retention after CNC incorporation introduced a favorable microenvironment for cell proliferation and differentiation [45].

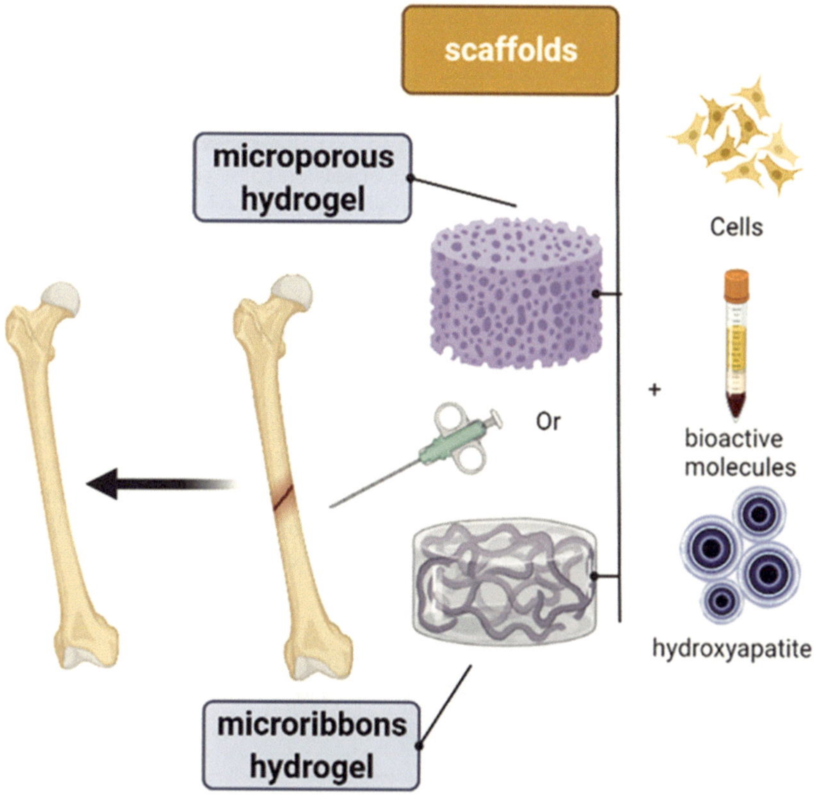

Figure 5.
Design of porous and hydrophilic scaffolds through functionalization of cotton cellulose with bioactive molecules (reproduced with copyright permission from [42], MDPI open access journal).

4. Composites in civil engineering: Additive effect of cotton waste on the macroscopic properties

The crystallinity and degree of polymerization of cotton cellulose are substantially higher than those of other agricultural residues such as rice straw and sugarcane bagasse. Moreover, the abundance of hydroxyl groups in cellulose introduces active surface sites for moisture absorption and hydrogen bonding interactions [1, 3, 4]. Thus, the advantages of waste cotton fiber enhance the mechanical properties of rigid materials used in building and construction applications. The specific outcomes of the cotton fiber reinforcement in an epoxy thermoset polymer were the augmentation of the flexural properties and equilibrium water content [46]. The compression molding process has been commonly applied to increase the interfacial adhesion of cotton fiber with the solid polymer matrix [47].

Combining waste polyester cotton with waste polyurethane by the hot-pressing technique introduced novel properties of flame retardance and sound absorption in the building materials. Moreover, the precision in the temperature (240°C) and pressure (12,000 psi) of the hot-pressing process improved the bending and impact strength of the multifunctional composite material. Further, the thermal and mechanical properties of the cotton-blended polyurethane composite were equivalent to those of wood and gypsum board, respectively [48]. In the case of bio-geopolymer preparation, the fiber-matrix bonding performance was evaluated between cotton stalks and cotton fibers. The alkali pretreatment and polyvinyl alcohol pretreatment of cotton stalk favored compact aggregation and stronger interplanar interaction between stalk and fiber. The aggregation was observed to be similar to the filling and cementing phenomena. As a result of improved bonding performance, the compressive and flexural strength were elevated by 5 and 11.5%, respectively [49].

A mixture of cotton textile waste and crushed bricks was applied as a replacement to natural sand in the making of concrete. In contrast to the multifunctional composite material described previously, the addition of cotton textile waste lowered the mechanical strength and density of concrete. However, the presence of cotton waste provided insulation properties and increased the water retention (**Figure 6**). Thus, the distinguishing feature of cotton waste blending included the production of faster curing and less thermally conductive concrete surfaces [51]. The use of industrial waste as a binder for concrete production has been reported to be more environmentally friendly as compared to the severely weathered bricks. Enlarged swelling behavior of cotton cellulose in sodium hydroxide solution widens the area of crystallization

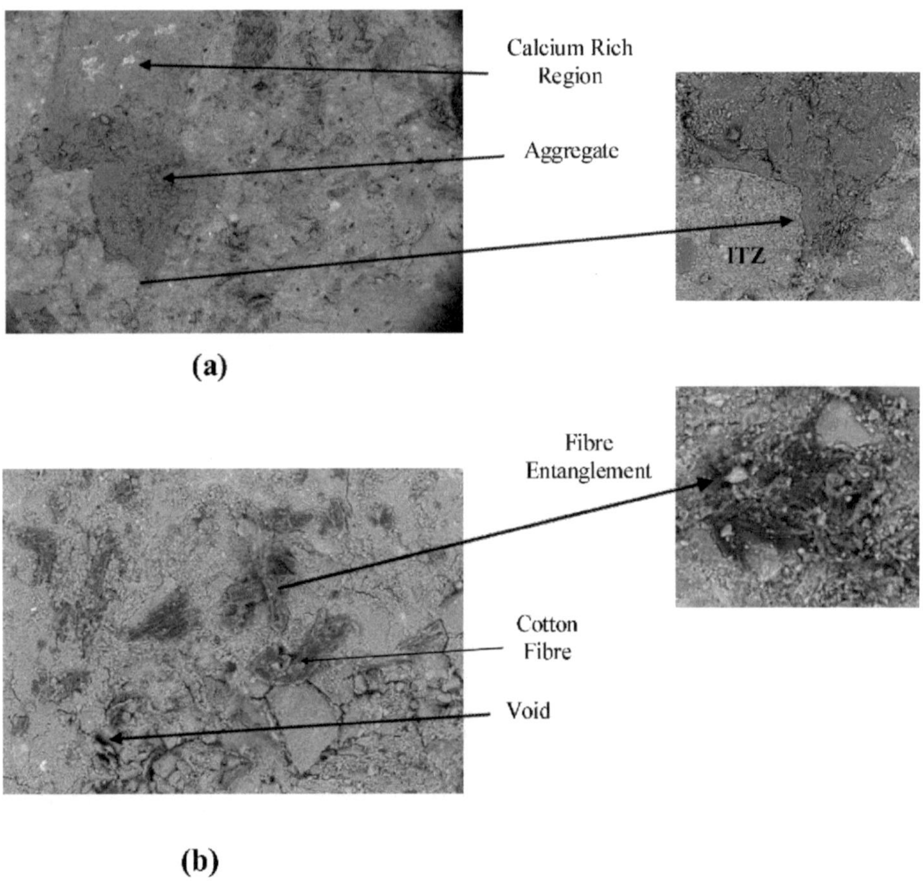

Figure 6.
Effect of cotton waste reinforcement and fiber entanglement on the curing process of concrete (reproduced with copyright permission from [50], MDPI open access journal).

and increases the compactness of concrete aggregation [52]. Thus, the incorporation of sodium hydroxide as an activator and cotton textile waste as a binder generated a thick and well-ordered concrete structure. For the concrete produced from black cotton soil, the use of 10 M sodium hydroxide at a constant activator-to-binder ratio of 0.35 enhanced the compressive strength. Besides, the industrial waste-derived concrete was reported to be associated with other environmental benefits such as lower carbon dioxide emissions and lower energy input [53].

The dyes present on the cotton textile waste cause a reduction in the interfacial adhesion between fiber and polyester and introduce asymmetric distribution of flexural strength and compressive properties across the fiberboard. The hot-pressing method has been reported to yield fiber composites of considerable compressive strength. The absorption of moisture, while it enhances the hardness of concrete material, negatively impacts the mechanical properties of fiberboard composites [54]. However, treating cotton waste with lime offers the dual advantages of removing dye and minimizing water retention, which supports the flexural and compressive properties of fiberboard composites [55]. In contrast to concrete structures, fiberboard composites and other lightweight composites require a relatively higher elastic modulus, flexural modulus, and compressive strength. In contrast, the mechanical resilience of concrete structures is measured by the material's hardness and impact strength [47, 51]. According to the field-specific application of the cotton waste reinforced composites, the composition of the reinforced cotton fiber varies. For instance, alkali pretreatment of waste cotton is indispensable for strengthening the solid polymer matrix formation in the concrete structures [53]. However, the blending of untreated cotton fibers and polyester is essential in the development of fiberboards (**Figure 7**). An exceptional scenario demands alkali pretreatment for constructing the fiberboards with polymers derived from the same plant source (e.g., cotton stalk and cotton fiber) [49].

Cotton stalks have been utilized to elevate the shelf life of wood biocomposites through the integration of the fungal cultivation method. The autoclave treatment

Figure 7.
Fabrication of all-cellulose composite fiberboard through interfacial adhesion between cotton waste-derived mechanical pulp fiber and cellulose nanofibril suspension (reproduced with copyright permission from [56], Elsevier Open access).

of cotton stalk increased its pore size and created amenable surface structures for the growth of *Aspergillus niger* mycelium. Interestingly, the cotton stalk acted as an activator, and the exopolysaccharide produced by the fungus served as the binder for producing the alternative wood composite. The additive effect of the cotton stalk on the wood board contributed to a rise in the thermal insulation property. Further, the coating of linseed oil from cottonseed mitigated the degradation of wood by termites and augmented the shelf life of the alternative wood composite panel [57]. The partial dissolution of cotton waste using sodium hydroxide/urea generates a chimeric composite with crystalline and amorphous regions. The all-cellulose composite, primarily produced from cotton textile waste, exhibited strong insulation and acoustic absorption properties. Importantly, the partial regeneration of cellulose II in the sodium hydroxide/urea mixture contributed to the crystallinity reduction and elevation of the noise reduction coefficient [58].

The macroscopic alterations in both concrete and fiberboard composites have been accurately measured by the Charpy test. The localized distribution of load-bearing units causes failure or fracture in the Charpy test. For instance, variations in the dispersion of cotton waste in the polyester and concrete structures with respect to unit square area of the composite modulated the impact strength between 10 and 15 KJ m^{-2} considerably. The disproportionate distribution changed the equilibrium. Hence, sudden displacement caused by the pendulum hammer introduced cracks and fractures depending on the deviation in the homogeneity of the composite [59, 60].

5. Discussion

Cotton waste, arising either from yarn production or post-consumer waste textiles, is unique with respect to the physiochemical properties of cellulose [1–4]. Cotton cellulose has been reported to show a high fiber length-to-diameter ratio, even after the mechanical fibrillation process. Though the semi-crystalline nature of cellulose is widespread in multiple plant biomasses, the major distribution of crystalline regions in cotton cellulose enhances the yield of cellulose nanocrystal formation by acid hydrolysis. Moreover, the preprocessing of cotton cellulose has lower chemical requirements due to the loose organization of non-cellulosic components [8–11]. Compared to regenerated cellulose from other agricultural crop residues, the cellulose from waste cotton has been observed to be associated with more thermostable and elastic properties [12, 13]. The ionic liquid treatment using NMMO has been extensively applied to derive viscose fibers, which exhibit increased reactivity to chemical functionalization with chitosan. The solvation of cotton cellulose with ionic liquids disrupts crystallinity and increases the porosity of the regenerated cellulose. Thus, the cellulose II arising after regeneration is amenable to efficient chemical modification, which widens the scope of multifunctional composite preparations against bacterial infections in chronic wounds [12, 35, 37]. On the other hand, the cellulose nanocrystals derived from cotton cellulose possess binding affinity to gelatin and egg white proteins in the tissue scaffold development. Though the cellulose lacks functional groups for cell adhesion and proliferation, the blending of either cellulose nanofibril or nanocrystal has been observed to be indispensable for introducing mechanical resilience. The flexibility of the scaffold in allowing tissue growth and regeneration has been correlated with the elevated compressive modulus factor of the cellulose matrix [38, 44, 45].

Apart from chemical treatments, cotton waste has been largely investigated for carbonization in anaerobic environments. As compared to other agricultural feedstocks, the thermochemical decomposition of cotton waste produces lower bio-oil and syngas [14]. The biochar production from cotton cellulose is significant, owing to the minimal weight loss at 450°C. Moreover, the retention of amorphous cellulose after thermochemical decomposition elevates the biochar's applicability in drug delivery systems [31, 32]. For instance, the hydrophilicity of the cotton cellulose-derived biochar facilitated the development of antibacterial fluorescent sensors [39]. On the contrary, the dissolution process using a sodium hydroxide/urea aqueous two-phase system integrates cellulose with iron oxide nanoparticles in the fabrication of a magnetic field-responsive drug delivery system. Moreover, partial cellulose dissolution at low temperatures using a sodium hydroxide/urea mixture significantly facilitates the production of an all-cellulose composite [40, 41, 58]. Lithium chloride, in the presence of N, N-dimethylacetamide, favors the construction of all-cellulose composite in the manufacture of fracture plaster [28]. Nevertheless, mechanical disintegration of cotton cellulose is the most commonly reported preprocessing step for versatile applications in the polymer composite preparations. The alteration of cellulose caused by the chemical treatments impedes the mechanical properties in cotton cellulose-reinforced fiberboard preparations [9, 24, 54]. The alkali bleaching enhances the porosity of the treated cellulose fiber and finds it suitable for modulating the curing process of concrete structures [53]. Though the moisture absorption properties of concrete and fiberboard are widely contrasting, the effective incorporation of cotton cellulose has been shown to cause magnificent elevation in the thermal resistivity, noise absorption coefficient, and flexural strength of building composite materials [47–49, 51]. Cotton stalk, owing to the much more compact association of non-cellulosic components, is molded into a standalone composite in the fabrication of a wood-like composite panel. The partial dissolution by the alkali urea mixture has been pivotal in the successful manufacture of acoustic-absorbing and heat-resistant cotton boards [58].

6. Conclusion

The chapter summarized the influences of various physical characteristics of cotton cellulose on the preparation of soft and hard composite materials. Soft materials were distinguished by significantly higher elastic and compressive moduli. Moreover, the microscopic distribution of polymer chains contributed to the development of multifunctional composites. For instance, the presence of aqueous media in the regenerated cotton cellulose aided in the transport of gases, nutrients, and therapeutic precursors for tissue engineering, wound healing, and drug delivery applications. On the contrary, the hard materials were characterized by the inter-adhesion between solid matrices. Unlike water-filled hydrogels, the polymeric composites for building and construction applications showed variable moisture retention. Factors such as bending strength, impact strength, and thermal stability were augmented through the cotton waste reinforcement in hard polymeric composites. Notably, the interfacial adhesion between chemically treated cotton stalk and mechanically disintegrated cotton waste fiber gave rise to a versatile class of fiberboards. Besides, the additive reinforcement using fungal exopolysaccharides and linseed oil led to the emergence of mechanical robustness on a macroscopic scale.

Acknowledgements

The author acknowledges VIT Bhopal University for providing access to full-length articles and review papers published in Scopus and SCI-indexed journals. No funding was received during the preparation of the article.

Conflict of interest

The author declares no conflict of interest.

Author details

Vignesh Natarajan
School of Biosciences Engineering and Technology, VIT Bhopal University, Bhopal, Madhya Pradesh, India

*Address all correspondence to: vignesh.n@vitbhopal.ac.in

References

[1] Reynoud N, Petit J, Bres C, Lahaye M, Rothan C, Marion D, et al. The complex architecture of plant cuticles and its relation to multiple biological functions. Frontiers in Plant Science. 2021;**12**:782773. DOI: 10.3389/fpls.2021.782773

[2] Sarkar D, Bu L, Jakes JE, Zieba JK, Kaufman ID, Crowley MF, et al. Diffusion in intact secondary cell wall models of plants at different equilibrium moisture content. The Cell Surface. 2023;**9**:100105. DOI: 10.1016/j.tcsw.2023.100105

[3] He Z, Nam S, Tewolde H, Ford CV, Dhandapani R, Barretto R, et al. Morphologic features and thermal characteristics of nine cotton biomass byproducts. Biomass. 2025;**5**(1):12. DOI: 10.3390/biomass5010012

[4] Pandirwar AP, Khadatkar A, Mehta CR, Majumdar G, Idapuganti R, Mageshwaran V, et al. Technological advancement in harvesting of cotton stalks to establish sustainable raw material supply chain for industrial applications: A review. BioEnergy Research. 2023;**16**(2):741-760. DOI: 10.1007/s12155-022-10520-3

[5] Chen C, Xi Y, Weng Y. Recent advances in cellulose-based hydrogels for tissue engineering applications. Polymers. 2022;**14**(16):3335. DOI: 10.3390/polym14163335

[6] Moradpoor H, Mohammadi H, Safaei M, Mozaffari HR, Sharifi R, Gorji P, et al. Recent advances on bacterial cellulose-based wound management: Promises and challenges. International Journal of Polymer Science. 2022;**2022**(1):1214734. DOI: 10.1155/2022/1214734

[7] Riaz S, Jabbar A, Siddiqui H, Salman M, Sarwar A. A sustainable process for cotton and polyester/cotton blend dyeing with nucleophilic disperse dyes through chemical modification. Cellulose. 2024;**31**(6):3981-3992. DOI: 10.1007/s10570-024-05820-0

[8] Banvillet G, Depres G, Belgacem N, Bras J. Alkaline treatment combined with enzymatic hydrolysis for efficient cellulose nanofibrils production. Carbohydrate Polymers. 2021;**255**:117383. DOI: 10.1016/j.carbpol.2020.117383

[9] Antony Jose S, Cowan N, Davidson M, Godina G, Smith I, Xin J, et al. A comprehensive review on cellulose nanofibers, nanomaterials, and composites: Manufacturing, properties, and applications. Nanomaterials. 2025;**15**(5):356. DOI: 10.3390/nano15050356

[10] Wang X, Chen P, Ogawa Y, Nishiyama Y, Qi H. Structural changes during heterogeneous sulfation and following homogenization of cotton cellulose. Cellulose. 2024;**31**(5):2729-2742. DOI: 10.1007/s10570-024-05787-y

[11] Haouache S, Jimenez-Saelices C, Cousin F, Falourd X, Pontoire B, Cahier K, et al. Cellulose nanocrystals from native and mercerized cotton. Cellulose. 2022;**29**(3):1567-1581. DOI: 10.1007/s10570-021-04313-8

[12] Zeng C, Wang H, Bu F, Cui Q, Li J, Liang Z, et al. A regenerated cellulose fiber with high mechanical properties for temperature-adaptive thermal management. International Journal of Biological Macromolecules. 2024;**274**:133550. DOI: 10.1016/j.ijbiomac.2024.133550

[13] Zhu M, Ying D, Zhang H, Xu X, Chang C. Self-healable hydrophobic films fabricated by incorporating natural wax into cellulose matrix. Chemical Engineering Journal. 2022;**446**:136791. DOI: 10.1016/j.cej.2022.136791

[14] Tujjohra F, Hoque E, Kader MA, Rahman MM. Sustainable valorization of textile industry cotton waste through pyrolysis for biochar production. Cleaner Chemical Engineering. 2025;**11**:100161. DOI: 10.1016/j.clce.2025.100161

[15] Primaz CT, Ribes-Greus A, Jacques RA. Valorization of cotton residues for production of bio-oil and engineered biochar. Energy. 2021;**235**:121363. DOI: 10.1016/j.energy.2021.121363

[16] Chojnacka K, Moustakas K, Mikulewicz M. Multifunctional cellulose-based biomaterials for dental applications: A sustainable approach to oral health and regeneration. Industrial Crops and Products. 2023;**203**:117142. DOI: 10.1016/j.indcrop.2023.117142

[17] Vassileva P, Uzunov I, Popova T, Voykova D, Avramova I, Mehandjiev D. Biochars as a solution for silver removal and antimicrobial activity in aqueous systems. Applied Sciences. 2025;**15**(5):2796. DOI: 10.3390/app15052796

[18] Abtew MA, Atalie D, Dejene BK. Recycling of cotton textile waste: Technological process, applications, and sustainability within a circular economy. Journal of Industrial Textiles. 2025;**55**:15280837251348663. DOI: 10.1177/15280837251348663

[19] Goyal A. Management of spinning and weaving wastes. In: Waste Management in the Fashion and Textile Industries. Cambridge, United States: Woodhead Publishing; 2021. pp. 61-82. DOI: 10.1016/B978-0-12-818758-6.00003-X

[20] Michelin M, Gomes DG, Romaní A, Polizeli MD, Teixeira JA. Nanocellulose production: Exploring the enzymatic route and residues of pulp and paper industry. Molecules. 2020;**25**(15):3411. DOI: 10.3390/molecules25153411

[21] Hongrattanavichit I, Aht-Ong D. Nanofibrillation and characterization of sugarcane bagasse agro-waste using water-based steam explosion and high-pressure homogenization. Journal of Cleaner Production. 2020;**277**:123471. DOI: 10.1016/j.jclepro.2020.123471

[22] Savadekar NR, Karande VS, Vigneshwaran N, Kadam PG, Mhaske ST. Preparation of cotton linter nanowhiskers by high-pressure homogenization process and its application in thermoplastic starch. Applied Nanoscience. 2015;**5**(3):281-290. DOI: 10.1007/s13204-014-0316-3

[23] Hemmati F, Jafari SM, Taheri RA. Optimization of homogenization-sonication technique for the production of cellulose nanocrystals from cotton linter. International Journal of Biological Macromolecules. 2019;**137**:374-381. DOI: 10.1016/j.ijbiomac.2019.06.241

[24] Wu Q, Ding C, Wang B, Rong L, Mao Z, Feng X. Green, chemical-free, and high-yielding extraction of nanocellulose from waste cotton fabric enabled by electron beam irradiation. International Journal of Biological Macromolecules. 2024;**267**:131461. DOI: 10.1016/j.ijbiomac.2024.131461

[25] Zhang L, Tan J, Xing G, Dou X, Guo X. Cotton stalk-derived hydrothermal carbon for methylene blue dye removal: Investigation of the raw material plant tissues. Bioresources and Bioprocessing. 2021;**8**(1):10. DOI: 10.1186/s40643-021-00364-8

[26] De Silva R, Byrne N. Utilization of cotton waste for regenerated cellulose fibres: Influence of degree of polymerization on mechanical properties. Carbohydrate Polymers. 2017;**174**:89-94. DOI: 10.1016/j.carbpol.2017.06.042

[27] Haule LV, Carr CM, Rigout M. Preparation and physical properties of regenerated cellulose fibres from cotton waste garments. Journal of Cleaner Production. 2016;**112**:4445-4451. DOI: 10.1016/j.jclepro.2015.08.086

[28] Kale RD, Gorade VG. Potential application of medical cotton waste for self-reinforced composite. International Journal of Biological Macromolecules. 2019;**124**:25-33. DOI: 10.1016/j.ijbiomac.2018.11.196

[29] Al Afif R, Pfeifer C, Pröll T. Bioenergy Recovery from Cotton Stalk. Faisalabad, Pakistan: IntechOpen; 2019. DOI: 10.5772/intechopen.88005

[30] Batuer A, Chen D, Xin Q, He X, Zhang J, Huang Z. Mechanical properties of waste cotton and their changes during early pyrolysis. Journal of Analytical and Applied Pyrolysis. 2020;**149**:104855. DOI: 10.1016/j.jaap.2020.104855

[31] Rubi RV, Allayban JP, Deduque DA, Mena JA, Robles NM, Rubio BD, et al. Production and characterization of activated carbon from pyrolysis biochar of cellulosic cotton-based textile wastes. Materials Today: Proceedings. Article in press. 2023. DOI: 10.1016/j.matpr.2023.05.443

[32] Çay A, Yanık J, Akduman Ç, Duman G, Ertaş H. Application of textile waste derived biochars onto cotton fabric for improved performance and functional properties. Journal of Cleaner Production. 2020;**251**:119664. DOI: 10.1016/j.jclepro.2019.119664

[33] Suresh N, Sivakumar P, Malathi AC, Balamurugan KS. Effect of temperature on EMI shielding behavior of jack fruit rags biochar and waste silk fiber-reinforced vinyl ester composite. Journal of Materials Science: Materials in Electronics. 2025;**36**(2):100. DOI: 10.1007/s10854-024-14161-4

[34] Kupnik K, Primožič M, Kokol V, Leitgeb M. Nanocellulose in drug delivery and antimicrobially active materials. Polymers. 2020;**12**(12):2825. DOI: 10.3390/polym12122825

[35] Mahmud RU, Alam R, Islam R, Hasan Z, Moniruzzaman M, Islam T. Chitosan-functionalized viscose fiber from post-consumer cotton waste for durable biomedical materials. Materials Research Express. 2025;**12**(2):025401. DOI: 10.1088/2053-1591/adb12f

[36] Waheed Z, Ahmad F, Mushtaq B, Ahmad S, Habib SR, Rasheed A, et al. Biowaste rice husk derived cellulosic hydrogel incorporating industrial cotton waste nonwoven for wound dressing. International Journal of Biological Macromolecules. 2024;**281**:136412. DOI: 10.1016/j.ijbiomac.2024.136412

[37] Salman SM, Azam F, Ahmad S, Ahmad F, Mushtaq B, Habib SR, et al. Development of environment friendly biodegradable biopolymer-based hydrogel composites incorporated with waste cotton for wound dressings. Journal of Industrial Textiles. 2025;**55**:15280837251324895. DOI: 10.1177/15280837251324895

[38] Hivechi A, Bahrami SH, Siegel RA. Drug release and biodegradability of electrospun cellulose nanocrystal reinforced polycaprolactone. Materials Science and Engineering: C. 2019;**94**:929-937. DOI: 10.1016/j.msec.2018.10.037

[39] Papadopoulou EL, Barbetta A, Fiorentini F, Lenzuni M, Carzino R,

Dante S, et al. Upcycling of cotton waste to functional carbon dots: Photocatalysis and antibacterial action under vis-NIR light. Surfaces and Interfaces. 2024;**48**:104241. DOI: 10.1016/j.surfin.2024.104241

[40] Chumpitaz D, Mariños Y, Peña E, Sencia JJ, Zapata A, Ponce S, Felix LL, Quispe LT, López R, Gutarra A. Sustainable magnetic hydrogels from cotton fabric waste as candidate for dual-modality drug release. First look Materials Today Family of Journals. 18 July 2025:32. Available from: https://ssrn.com/abstract=5357318

[41] Azizi A. Green synthesis of Fe3O4 nanoparticles and its application in preparation of Fe3O4/cellulose magnetic nanocomposite: A suitable proposal for drug delivery systems. Journal of Inorganic and Organometallic Polymers and Materials. 2020;**30**(9):3552-3561. DOI: 10.1007/s10904-020-01500-1

[42] Alkhursani SA, Ghobashy MM, Al-Gahtany SA, Meganid AS, Abd El-Halim SM, Ahmad Z, et al. Application of nano-inspired scaffolds-based biopolymer hydrogel for bone and periodontal tissue regeneration. Polymers. 2022;**14**(18):3791. DOI: 10.3390/polym14183791

[43] Singh S, Bhushan S, Khan H, Chaudhari LR, Ali A, Das A, et al. Surgical cotton microfibers loaded with proteins and apatite: A potential platform for bone tissue engineering. International Journal of Biological Macromolecules. 2023;**236**:123812. DOI: 10.1016/j.ijbiomac.2023.123812

[44] Ahmed MZ, Mutahir S, Intisar A, Muhammad N, Mutahir Z. Probe sonicated cotton cellulose fibers treated with bone conditioned medium for optimum cell growth and mineralization. Cellulose. 2023;**30**(12):7497-7518. DOI: 10.1007/s10570-023-05386-3

[45] Rashtchian M, Hivechi A, Bahrami SH, Milan PB, Simorgh S. Fabricating alginate/poly (caprolactone) nanofibers with enhanced bio-mechanical properties via cellulose nanocrystal incorporation. Carbohydrate Polymers. 2020;**233**:115873. DOI: 10.1016/j.carbpol.2020.115873

[46] Kamble Z, Behera BK. Fabrication and performance evaluation of waste cotton and polyester fiber-reinforced green composites for building and construction applications. Polymer Composites. 2021;**42**(6):3025-3037. DOI: 10.1002/pc.26036

[47] Wankhede B, Bisaria H, Ojha S, Dakre VS. A review on cotton fibre-reinforced polymer composites and their applications. Proceedings of the Institution of Mechanical Engineers, Part L: Journal of Materials: Design and Applications. 2023;**237**(6):1347-1362. DOI: 10.1177/14644207221143876

[48] Liang F, Chen Q, Shi Z, Dan R, Liang H, Feng X, et al. Multifunctional composites made from waste polyester cotton fabric and waste rigid polyurethane foam for potential building wall material applications. Energy and Buildings. 2025;**329**:115274. DOI: 10.1016/j.enbuild.2025.115274

[49] Zhou B, Wang L, Ma G, Zhao X, Zhao X. Preparation and properties of bio-geopolymer composites with waste cotton stalk materials. Journal of Cleaner Production. 2020;**245**:118842. DOI: 10.1016/j.jclepro.2019.118842

[50] Tang W, Monaghan R, Sajjad U. Investigation of physical and mechanical properties of cement mortar incorporating waste cotton fibres. Sustainability. 2023;**15**(11):8779. DOI: 10.3390/su15118779

[51] Saca N, Radu L, Truşcă R, Calotă R, Dobre D, Năstase I. Assessment of properties and microstructure of concrete with cotton textile waste and crushed bricks. Materials. 2023;**16**(20):6807. DOI: 10.3390/ma16206807

[52] Teklehaimanot M, Hailay H, Tesfaye T. Manufacturing of ecofriendly bricks using microdust cotton waste. Journal of Engineering. 2021;**2021**(1):8815965. DOI: 10.1155/2021/8815965

[53] Gaddam AG, Gondu VR, Yamsani SK, Adigopula VK. Comprehensive utilization of brick waste as a precursor in synthesizing geopolymer for treating black cotton soil. International Journal of Geotechnical Engineering. 2024;**18**(4):360-366. DOI: 10.1080/19386362.2023.2245230

[54] Liang F, Shi Z, Dan R, Liang H, Feng X, Shang X, et al. The influence of dyes on the mechanical properties of composites fabricated from waste polyester/cotton fibers and their potential application value in the construction field. Composites Part B: Engineering. 2025;**303**:112576. DOI: 10.1016/j.compositesb.2025.112576

[55] Gautam KK, Patel S. An experimental investigation to stabilize black cotton soil using lime, waste quarry dust, and polyester fiber. In: Indian Geotechnical Conference. Singapore: Springer Nature Singapore; 2024. pp. 295-312. DOI: 10.1007/978-981-96-2444-7_23

[56] Hossain R, Tajvidi M. Cellulose nanofibrils as a bio-based binder for wood fiber composite insulation panels with enhanced thermo-mechanical properties for structural wall sheathing applications. Carbohydrate Polymer Technologies and Applications. 2024;**7**:100429. DOI: 10.1016/j.carpta.2024.100429

[57] Ibrahim RA, Fahim IS, Shaban M, Gomaa OM. Sustainable wood composite production using cotton waste and exopolysaccharides as green binders. International Journal of Biological Macromolecules. 2024;**281**:135710. DOI: 10.1016/j.ijbiomac.2024.135710

[58] Suthatho A, Rattanawongkun P, Tawichai N, Tanpichai S, Boonmahitthisud A, Soykeabkaew N. Low-density all-cellulose composites made from cotton textile waste with promising thermal insulation and acoustic absorption properties. ACS Applied Polymer Materials. 2023;**6**(1):390-397. DOI: 10.1021/acsapm.3c02076

[59] Manikam B, Thangaraju J, Manoharan M, Admassu Y. Experimental analysis and development of sustainable composites from waste cotton/coffee husk/sawdust for thermal insulating materials. Journal of Engineered Fibers and Fabrics. 2025;**20**:15589250251342869. DOI: 10.1177/15589250251342869

[60] Baccouch W, Ghith A, Yalcin-Enis I, Sezgin H, Miled W, Legrand X, et al. Investigation of the mechanical, thermal, and acoustical behaviors of cotton, polyester, and cotton/polyester nonwoven wastes reinforced epoxy composites. Journal of Industrial Textiles. 2022;**51**(6):876-899. DOI: 10.1177/1528083720901864

Chapter 6

Enhancement of Sustainability, Inclusiveness, and Value-Adding Capabilities of the Cotton Value Chain through Collaborative Sustainability-Based Innovation between Italy and Egypt

Dalia Gallico

Abstract

This chapter examines the critical transformation needed in Egypt's cotton-textile-garment (CTG) value chain to transition from a commodity-based economy to one driven by value addition, sustainability, and competitiveness. Despite its historical reputation for producing extra-long staple (ELS) cotton of exceptional quality, Egypt has experienced a significant decline in both cotton production and industrial processing due to outdated infrastructure, weak integration between agricultural and industrial sectors, and inadequate human capital development. Drawing on best practices from national and international case studies, including the Cottonforlife initiative, this study proposes an integrated model for revitalizing the Egyptian cotton sector. The proposed framework integrates sustainable farming methods, including organic cultivation and integrated pest management, with technical enhancements, resource-efficient production, and institutional strengthening. Special emphasis is placed on enabling local value addition through spinning, dyeing, and finishing activities, as well as equipping small and medium-sized enterprises (SMEs), start-ups, and vocational institutions with the capacity to meet global market standards. The chapter also explores strategies for enhancing quality assurance, traceability, and compliance with international standards, while fostering public-private partnerships and industrial clustering to improve productivity and innovation. By integrating environmental, economic, and social dimensions, this model aims to reposition Egyptian cotton as a premium product in domestic and global markets. The study concludes that sustainable transformation of the CTG value chain requires coordinated investments in technology, skills development, market access, and policy reform.

Keywords: Egyptian cotton, sustainable development, sustainable agriculture, education, active involvement, public-private partnerships, organic cotton production, industrial upgrading

1. Introduction

Cotton remains one of the most historically significant and economically vital fiber crops globally. Cultivated across over 100 nations, it occupies approximately 2.5% of the world's arable land—an estimated 35 million hectares—positioning it just behind staple crops such as cereals and soybeans in terms of land usage. In parallel, cotton is a major globally traded commodity, involving over 150 countries in its import and export dynamics [1].

Its economic relevance is particularly significant for low- and middle-income countries, where cotton cultivation and processing play a key role in driving employment and rural development. The cotton value chain is complex and multi-staged, comprising agricultural research and innovation, field cultivation, fiber harvesting, initial processing for commercialization, conversion into yarn, fabric production, finishing, and eventual transformation into garments and household products. An estimated 100 million rural households—predominantly in developing regions—are directly involved in cotton farming. When ancillary activities, such as transportation, ginning, baling, and warehousing, are included, the number of people reliant on the cotton industry exceeds 350 million. Furthermore, the global textile and apparel sectors alone provide employment to an estimated 60–75 million individuals [2].

Historically, the cotton and textile industries have been a major player in Egypt's national economy. Egyptian cotton—specifically the extra-long staple (ELS) variety *Gossypium barbadense*, colloquially referred to as "White Gold"—is globally renowned for its exceptional softness, length, and durability. Grown in the fertile, humid conditions of the Nile Delta, Egyptian cotton fibers are handpicked to preserve quality, ensuring a higher degree of fiber uniformity and strength. The result is a luxury-grade textile highly sought after by premium international brands.

From an economic standpoint, Egypt's textile sector makes up around 3% of the country's gross domestic product (GDP), accounts for a third of manufacturing value-added, employs roughly a third of the industrial labour force, and produces 15% of Egypt's non-oil exports—amounting to about $2.6 billion annually [3].

Nevertheless, despite its historical prominence, Egypt's role in the contemporary global textile value chain remains limited. As identified in the national strategy document Vision 2025: Expansion & Employment in the Textile Sector, the industry is characterized by fragmentation, outdated practices, and insufficient integration with international quality standards.

Several structural issues persist, including underinvestment in modernization, high production costs, a shortage of skilled labour, and limited adoption of sustainable and innovative agricultural techniques. The textile production process is primarily dominated by spinning and weaving operations, which are largely operated by state-owned enterprises. While private-sector participation is slowly increasing, public ownership has led to inefficiencies in technological upgrades, responsiveness to consumer demand, and financial performance. The weakest links in the value chain remain dyeing and finishing, which receive the least investment and innovation.

As a consequence of these systemic constraints, Egyptian textile manufacturers increasingly depend on imported intermediate goods. Yarns and finished fabrics are frequently sourced from textile-exporting countries such as India, Turkey, Bangladesh, and Pakistan to meet the domestic demand for ready-made garments. In some cases, Egyptian cotton is exported in its raw form, processed abroad, and then re-imported as finished textile products [4].

Egypt primarily exports long-staple (LS) and extra-long-staple (ELS) raw cotton. However, over half of its textile exports—approximately 52.8%—are in the form of garments manufactured using imported yarns. Another 25.4% comprises finished fabrics and home textiles made from locally sourced raw materials. The majority of raw cotton exports are directed towards low-cost manufacturing hubs, where they are processed and transformed into high-value products, such as garments branded as 100% Egyptian cotton but produced outside Egypt.

State-owned firms have been unable to effectively compete with lower-cost imported goods, leading to the decline or closure of several production facilities. While Egypt experienced steady growth in both cotton cultivation and textile production between the 1940s and 1970s, the sector began to contract significantly from the 1980s onwards. The shift in consumer preference towards more affordable synthetic fibers, particularly polyester, further contributed to the decline in domestic sales and international competitiveness. As marketing challenges persisted, many Egyptian farmers abandoned cotton cultivation in favor of crops offering higher and more stable returns.

The continued rise in imports—including textiles, dyes, and auxiliary materials—combined with a stagnant export sector, has led to a negative trade balance. Rather than serving as a foreign exchange earner, Egypt's textile industry has increasingly become a net consumer of financial resources [5].

2. Cotton production dynamics in Egypt

The contraction in Egypt's cotton output over recent decades is closely associated with a sustained decline in foreign demand, particularly for lint cotton. A key factor contributing to this downturn is the perceived deterioration in product quality, especially concerning seed purity and lint contamination. Pre- and post-harvest operations, including manual picking techniques, handling procedures, storage conditions, transportation logistics, and ginning practices, critically influence these quality parameters. Among these, contamination by non-vegetal materials—notably in handpicked cotton—has emerged as a primary constraint. Such contamination not only undermines the esthetic and structural integrity of the lint but also induces complications during the spinning process, prompting significant price discounts to offset the additional processing costs borne by textile manufacturers [6].

Despite possessing favorable agro-ecological conditions ideally suited for low-input cultivation, organic cotton production in Egypt remains negligible. This limited engagement in organic farming contrasts with the rapidly growing global demand for sustainable raw materials. In recent years, numerous multinational fashion and textile brands have committed to adopting 100% organic cotton within defined timelines, often by 2020, as part of broader sustainability pledges. Egypt's underutilization of this opportunity reflects a fragmented supply chain and insufficient infrastructural support for sustainable agriculture. To capitalize on emerging market trends, Egypt must adopt a systemic approach to supply chain integration, starting with certified organic seed development and traceable fiber production, and extending to environmentally responsible ginning and manufacturing practices. Moreover, promoting hybrid business models—which foster partnerships among smallholder farmers, cooperative mills, international retailers, and local innovators—could significantly enhance the country's competitiveness in the organic segment [7].

In response to these challenges, the Egyptian government has introduced regulatory interventions aimed at rehabilitating the country's reputation in international cotton markets. One such initiative is the Ministerial Decree No. 1918/2015, which reasserts public-sector control over seed qualification, production, and distribution. Under this framework, the government reinstated full ownership of seed operations and established a traceability system managed by the Cotton Egypt Association (CEA), encompassing all licensed handlers of Egyptian cotton across the value chain.

As part of the reform process, the Cotton Research Institute (CRI), operating under the auspices of the Ministry of Agriculture and Land Reclamation (MALR), implemented a targeted seed cultivation programme. High-quality cotton seeds were planted across 12,600 hectares in eight different governorates, yielding 41,237 cotton bales. Building on this initiative, MALR has signaled its intention to undertake additional seed purification cycles aimed at improving varietal consistency and fiber strength. It is projected that approximately 60% of the resulting seeds will be deployed in the upcoming 2019/2020 planting season [8].

Nevertheless, despite these strategic efforts and the anticipated improvements in production systems, extra-long staple (ELS) cotton output is expected to continue its downward trajectory. This persistent decline underscores the structural issues affecting the sector and highlights the need for deeper reform—not only in agronomic practices but also in market positioning, branding, and international collaboration.

Fammi sapere se vuoi che proceda con il Capitolo 3, oppure se desideri inserire:

3. Industrial transformation of the cotton sector in Egypt

Egyptian cotton—particularly its extra-long staple (ELS) varieties—has long been internationally acclaimed for its superior fiber length and softness, the capacity to fully exploit its inherent value remains constrained by technological and systemic limitations. Optimal processing of ELS cotton requires advanced, high-precision machinery capable of preserving fiber integrity during spinning and weaving. However, the majority of Egypt's textile production facilities continue to operate with obsolete and inefficient equipment, which is largely inadequate for high-quality ELS fiber processing. As a result, domestic textile mills are unable to meet their production requirements solely with local cotton. Projections for the 2019/2020 season indicated a national lint cotton consumption of approximately 590,000 bales, of which nearly 70% were anticipated to be met through cotton imports, with only the remaining 30% sourced domestically [9].

This technological lag is compounded by a shortage of skilled human capital, which continues to undermine Egypt's competitiveness in the global textile value chain. Despite a series of technical and vocational education reforms—most notably under the European Union's Technical and Vocational Education and Training (TVET) I and II programmes—a structural mismatch persists between the outputs of the educational system and the specific demands of the labour market. According to the World Economic Forum's (WEO's) Global Competitiveness Report (2015–2016), Egypt ranked 131st out of 140 countries in terms of the overall quality of its educational system, with similarly low scores for science, mathematics, and management education, as well as for innovation capacity [10].

The labour shortfall is particularly acute in industrial and agro-industrial sectors, where the lack of alignment between agricultural training and industrial skill development has resulted in fragmented knowledge, sectoral silos, and limited understanding of the cotton value chain as a unified system. The absence of horizontal integration between educational institutions and private industry has constrained opportunities for innovation, productivity enhancement, and value addition.

DOI: http://dx.doi.org/10.5772/intechopen.1011629

To address these multifaceted challenges, the Egyptian Ministry of Trade and Industry (MTI), building on its long-standing cooperation with the United Nations Industrial Development Organization (UNIDO), has sought to expand technical support for modernizing the national textile value chain. This initiative aligns with UNIDO's broader development mandate, which emphasizes inclusive industrialization, environmental sustainability, and poverty alleviation.

In this context, UNIDO proposed the implementation of a comprehensive development project entitled "From Cotton Seeds to Clothing: Enhancing the Sustainability, Inclusiveness, and Value Addition of the Cotton Value Chain in Egypt." The programme aims to improve the economic viability, environmental resilience, and social inclusiveness of both upstream and downstream actors in the cotton sector, while also strengthening institutional capacities, governance mechanisms, and market integration [11].

The design of this intervention draws upon the experiences and achievements of the Cottonforlife Initiative, a corporate social responsibility (CSR) programme initiated by FILMAR S.p.A. in partnership with AlexBank. Cottonforlife is a multi-stakeholder, five-year initiative focused on promoting sustainable fashion through the transparent, ethical, and eco-conscious production of textiles made from Egyptian cotton. Key milestones include the introduction of organic cotton cultivation (notably Giza 45 and Giza 87 varieties), supported through tripartite agreements among private-sector actors, farmers' associations, and the Ministries of Agriculture and Education [12].

One of Cottonforlife's flagship outputs is the production of NILO, an ultra-fine organic yarn that has gained recognition in prominent international textile exhibitions and attracted interest from globally renowned Italian fashion brands. Beyond product innovation, the initiative has also prioritized human capital development. Leveraging co-funding from AlexBank, the programme has been engaged in the modernization of agricultural and textile education, facilitating curriculum reform, teacher training, and the publication of updated instructional materials tailored to three-year technical school tracks. These actions aim to bridge the systemic gap between agricultural and industrial training, thus enabling more coherent skill formation along the entire cotton value chain.

4. Primary beneficiaries and target stakeholders

The proposed initiative aims to test and scale innovative, sustainable solutions that address persistent bottlenecks in the long-staple (LS) and extra-long staple (ELS) cotton value chains in Egypt. The project adopts a systemic and inclusive approach, targeting a minimum of 1000 direct beneficiaries drawn from key stakeholder categories integral to the sector's transformation.

The primary target groups include:

Cotton producers and associated farmer cooperatives will gain access to knowledge exchange platforms and structured capacity-building activities focused on sustainable agricultural techniques, contamination prevention, and organic cultivation practices. Enhanced market linkages and value chain integration are expected outcomes for this group.

Textile companies in the private sector, including small and medium-sized enterprises (SMEs) and start-ups led by young people, are involved in various downstream activities like spinning, weaving, knitting, dyeing, finishing, and making clothing.

The project aims to boost the productivity, environmental performance, social impact, and readiness for export of these companies.

Traders and retail intermediaries will benefit from greater transparency in product certification, improved traceability, and better coordination among local and international supply chain participants.

Workers in agriculture, technical staff, and textile professionals will get their skills upgraded through a mix of updated training programmes, hands-on technical training, and job placements that match industry needs.

Students in vocational and technical education programmes, especially those in secondary-level agricultural and industrial schools or vocational training, will get targeted support through updated learning paths aimed at closing the gap between education and industry demands.

Key stakeholders from institutions, such as technical support groups and specialized training centres, will take part in knowledge sharing and institutional development activities. The goal is to enhance service delivery, governance, and coordination across sectors.

A central pillar of the project's design is its emphasis on inclusive and equitable development, ensuring the proactive participation of underrepresented and vulnerable groups, including low-income populations, women, youth, and rural communities. The project aims not only to foster environmentally responsible agricultural environments—characterized by non-toxic inputs and biodiversity—but also to generate sector-specific entrepreneurial opportunities and pathways to employment across domestic and international markets.

Additionally, end consumers, both within Egypt and abroad, are positioned as indirect beneficiaries. By supporting quality assurance, sustainability, and traceability in the cotton production and processing chains, the project seeks to restore and enhance the reputation of Egyptian cotton, reinforcing its global competitiveness and contributing to increased consumer trust and demand.

5. Stakeholders' landscape and institutional engagement

In alignment with the mandate of the United Nations Industrial Development Organization (UNIDO) and its systemic approach to inclusive industrial development, the project will adopt a multi-stakeholder governance model. This model builds upon the institutional networks and collaborative mechanisms previously established under the ongoing CSR-driven initiative Cottonforlife, and aims to expand engagement across key public, private, and civil society actors in the cotton value chain [13].

At the heart of the project's architecture is the active involvement of the Ministry of Agriculture and Land Reclamation (MALR) and the Ministry of Trade and Industry (MTI), along with their partner research institutions and sectoral organizations. These key players are expected to play a crucial role in promoting sustainable cotton farming and industrial transformation, as well as shaping a regulatory framework that supports the growth of an environmentally sustainable and socially responsible textile sector.

Key government and research institutions:

Agricultural Research Center (ARC)—Founded in 1970, ARC's institutional mission is to maximize land and water productivity through innovation in agronomic research and resource optimization.

Cotton Research Institute (CRI)—Operating within ARC, the CRI adopts a multidisciplinary framework to foster synergies among farmers, agronomists, and technologists. Its strategic priorities include: (i) developing new cotton cultivars with enhanced yield potential and resistance to pests and climate stressors; (ii) safeguarding genetic seed purity; (iii) improving ginning techniques and fiber classification methods; and (iv) identifying optimal agro-climatic zones for enhanced fiber spinnability.

Central Laboratory of Organic Agriculture (CLOA)—CLOA promotes organic cultivation practices aligned with international standards. Its mandate includes pesticide reduction, certification and inspection services, and farmer capacity-building in sustainable agriculture.

Industry and trade institutions:

The project will leverage Egypt's industrial support infrastructure through engagement with MTI-affiliated bodies, including:

Egyptian Council for Textile Industries.

Industrial Modernization Center (IMC).

Federation of Egyptian Industries (FEI) and its Textile Chamber.

Textile Export Council (TEC) and Home Textile Export Council (HTEC).

These entities function as platforms for industrial coordination, trade facilitation, and policy dialog, particularly between domestic producers and international partners in the spinning, weaving, dyeing, and finishing sectors. The initiative will also integrate the expertise of the Textile Technology Center (TTC), Egyptian Organization for Standardization and Quality (EOS), Productivity and Vocational Training Department (PVTD), and the Fashion Design Center (FDC)—the latter established through a bilateral partnership with the Istituto di Moda Burgo (Milan). FDC offers both diploma-level and short-term fashion and design training aligned with global standards, while also providing incubation services and creative consultancy to emerging designers and small and medium-sized enterprises (SMEs).

Additional technical partners include:

Egyptian National Cleaner Production Center (ENCPC)—Focused on resource efficiency, renewable energy, and environmentally responsible manufacturing.

Productivity and Quality Enhancement Center (Kaizen)—Specializing in the application of lean manufacturing principles to enhance industrial productivity and quality control.

National and International Collaborations:

Through the existing network of Cottonforlife and its lead corporate sponsor, Filmar S.p.A., the project will foster cooperation with Italian institutions and industry bodies, including textile machinery manufacturers, fashion councils, and chambers of commerce (e.g., Unioncamere, Unionfiliere), to facilitate transnational knowledge exchange. Areas of collaboration will include textile standardization, technological innovation, sustainable product design, traceability, and global market access. The strategic partnership with AlexBank will continue to provide financial support mechanisms, particularly for start-ups and youth-led initiatives.

Regulatory and Certification Bodies:

Cotton Arbitration and Testing General Organization (CATGO)—An independent authority providing technical, arbitration, and certification services across the cotton value chain. CATGO ensures varietal purity, supervises quality control from field to export, and delivers accredited training in classing, testing, and moisture regulation.

Alexandria Cotton Exporters Association (ALCOTEXA) —A mandatory-membership organization for cotton traders and ginners, ALCOTEXA sets indicative export pricing and regulates international trade protocols.

Cotton Egypt Association (CEA) – Established by MTI and ALCOTEXA, CEA is the sole custodian of the Egyptian Cotton™ trademark. It employs DNA-based verification methods to authenticate product origin and prevent counterfeit branding.

Education and Training Partners:

The Ministry of Education will be directly involved in implementing project components related to technical and vocational education (TVET), particularly within secondary-level agricultural and industrial institutions. Collaborative arrangements will also be promoted with universities, such as the University of Alexandria, and technical institutes, including the Don Bosco Institute, as well as other international academic organizations with expertise in textile engineering, fashion design, and sustainable production curricula.

6. Strategic rationale and sectoral justification

Despite Egypt's global recognition for its high-quality extra-long staple (ELS) cotton, the domestic value-added potential remains largely underexploited. A substantial proportion of Egyptian cotton is exported in its raw form as a low-value commodity, rather than being processed locally. In several textile products manufactured in the country, imported components account for up to 70% of the inputs due to the underdevelopment of domestic supply and processing industries.

The transformation of raw cotton into yarn constitutes the first industrial conversion step and represents a critical inflection point in the value chain. Spinning significantly increases the economic value of cotton and acts as a gateway to industrial scalability. Unlike finished garments, cotton yarn occupies a stable and strategic position in global trade as an intermediate good, less influenced by rapidly shifting fashion trends and more resilient to market volatility [14].

The international cotton yarn market involves more than 30 participating countries and is valued at approximately USD 8 billion annually. The product's long shelf life further enhances its commercial viability. Importantly, China—the world's largest yarn importer—represents a unique market opportunity, as it exerts lower competitive pressure on other yarn-producing nations compared to its dominance in the fabric and garment segments [15].

From a development economics perspective, promoting yarn production and spinning infrastructure offers a practical pathway for Egypt to enhance its integration into global value chains (GVCs). This would also contribute to import substitution, job creation, and increased export potential.

In recognition of these opportunities, Egypt's national textile strategy, outlined in Vision 2025: Expansion & Employment, articulates a series of policy objectives and investment priorities aimed at transforming the textile and garment sector into a globally competitive industry. The strategy sets out ambitious targets, including:

Creation of one million new jobs, targeting both skilled and semi-skilled workers;

Training of over 500,000 personnel, with an additional 100,000 technical managers receiving advanced engineering diplomas;

An increase in textile exports to USD 10 billion, with a minimum of 70% locally sourced inputs;

Mobilization of over USD 13.5 billion in sectoral investment [16].

To achieve these outcomes, the strategy is structured around several core pillars:

Industrial base expansion and improved competitiveness of domestic firms;

Backward integration, including investments in fiber and raw textile production;

Export promotion and market diversification;

Human capital development, with emphasis on skill formation and entrepreneurship;

Policy reform and infrastructure investment to support sectoral modernization.

Global shifts in textile manufacturing and consumer expectations further underscore the urgency of this transformation. Sustainability, transparency, and traceability are no longer optional; they are preconditions for market access. This necessitates the adoption of new business models grounded in vertical integration and collaborative value chain governance. Strategic alliances among farmers, processors, textile brands—both large and small—are essential to ensure alignment with evolving standards for environmental and social responsibility [17].

The continued decline in Egyptian cotton production—driven by reduced demand and competitiveness—mirrors the contraction of the domestic textile industry in both volume and export capacity. This trend calls for a paradigm shift towards advanced, inclusive, and sustainable production systems. Enhancing sectoral resilience requires coordinated interventions to attract private investment, modernize equipment, and upgrade the labour force through targeted skill development.

Within this context, the present project seeks to:

Boost local production capabilities to enhance productivity, competitiveness, and sustainability;

Encourage public-private partnerships and business connections throughout the value chain;

Promote innovation and develop human capital;

Ultimately, rebrand Egyptian cotton as a high-value, traceable, and market-driven product in both domestic and international markets.

7. Comparative advantage and strategic positioning

The proposed project capitalizes on UNIDO's extensive technical expertise and operational track record in strengthening agro-industrial value chains, particularly within Egypt and across other cotton-producing regions. UNIDO has previously led successful interventions targeting the modernization of cotton-textile-garment (CTG) value chains across diverse geographic contexts—including South Asia, Latin America, West and Central Africa (in partnership with the World Trade Organization (WTO)), and the South Mediterranean—focusing on quality infrastructure, market access, processing technologies, and certification systems.

In addition, UNIDO's experience in resource-efficient and cleaner production (RECP) and its promotion of renewable energy applications in Egyptian industries provide the project with a robust foundation for embedding sustainability into all phases of cotton cultivation and processing. These existing frameworks will be adapted and scaled within the cotton value chain to foster eco-efficiency and industrial decarbonization.

Aligned with the 2030 Agenda for Sustainable Development, the project adopts a multi-stakeholder, inclusive approach that reflects Sustainable Development Goal (SDG) principles. By involving a range of actors—government agencies, private-sector investors, corporate social responsibility (CSR) partners, industry mentors, and non-traditional financiers—the initiative seeks to foster systemic change in Egypt's textile sector, improve cotton's contribution to GDP and exports, and expand employment, particularly among rural and youth populations.

The intervention strategy is built upon two complementary pillars:

Enhancement of the productivity, sustainability, and economic resilience of cotton growers, with an emphasis on long-staple (LS) and extra-long staple (ELS) varieties;

Improvement of the operational and competitive performance of private-sector textile enterprises, including SMEs and youth-led start-ups, particularly in processing, spinning, and value-added manufacturing.

These pillars reflect UNIDO's agri-value chain development methodology, which emphasizes the generation of local content, equitable value distribution, and environmental sustainability. This model ensures that all stages of the value chain—from input supply, cultivation, and fiber extraction to spinning, finishing, and marketing—are examined and optimized for value addition and market alignment. Target segments are selected based on their proven potential for economic return, social inclusion, and environmental benefit.

As part of its capacity-building agenda, the project will work to reinforce the institutional and regulatory architecture that governs cotton production and processing in Egypt. It will support technical bodies and regulatory agencies to develop policies, monitoring tools, and legal frameworks that facilitate the transition to an eco-conscious and socially accountable cotton industry.

Through these targeted interventions and in close collaboration with public and private stakeholders, UNIDO aims to achieve the following key outcomes:

Directly and indirectly benefit a minimum of 1000 individuals involved in the Egyptian cotton value chain;

Enhance access to employment and entrepreneurial pathways, particularly for youth, women, and rural workers;

Boost the productivity and global competitiveness of cotton growers, ginneries, and textile processing firms;

Stimulate growth in export earnings, reinforcing Egypt's positioning in high-value textile markets;

Facilitate domestic and foreign investment in cotton transformation industries;

Increase value addition at the national level, strengthening industrial linkages and backward integration;

Promote resource efficiency and sustainability in all phases of cotton production and processing, reducing waste and energy intensity.

By building on these comparative advantages, the project positions Egypt to regain and consolidate its place as a leader in high-quality cotton production, while embedding sustainability, inclusiveness, and innovation throughout the value chain.

8. Cotton sustainable production: Practices, capacity-building and quality assurance

Building on practical insights from ongoing CSR initiatives—most notably Filmar's operations in Egypt—and international benchmarks in sustainable agriculture, this project aims to transform LS and ELS cotton cultivation practices by promoting ecological integrity, resource efficiency, and market-driven quality standards. To achieve this, the initiative will implement pilot demonstration programmes that showcase agronomic innovations in organic farming, contamination mitigation, and the optimized use of water, fertilizers, and pest control agents. These demonstrations will be paired with technical training, field-based learning, and study tours, providing growers with both theoretical knowledge and practical exposure to end-market expectations, particularly from downstream actors such as spinning mills [18].

Promoting organic and low-impact cotton cultivation [19].

Organic cotton cultivation—based on the use of natural inputs and non-synthetic agrochemicals—enhances soil regeneration, avoids water and land contamination, and promotes farmer health and biodiversity conservation. The project aims to discourage reliance on conventional inputs, particularly herbicides and synthetic pesticides, which are associated with deteriorating soil quality, groundwater contamination, pest resistance, and ecological imbalance. Globally, pest-related damage accounts for approximately 15% of annual cotton losses, often leading farmers to overuse chemical treatments.

To counter this, the project will promote Integrated Pest Management (IPM) strategies and alternative pest control techniques, emphasizing non-chemical interventions to reduce toxic exposure and environmental degradation. Farmers will also be educated on the health risks associated with hazardous chemical handling and the importance of protective equipment, ensuring occupational safety [20].

Soil, water, and nutrient management.

To address the sector's water footprint, efficient irrigation systems and water stewardship practices will be encouraged to reduce wastage and enhance per-unit productivity. Simultaneously, soil fertility will be managed through precision nutrient application, aligned with site-specific crop requirements. These strategies aim to preserve soil structure, mitigate erosion, and ensure long-term productivity.

Byproduct utilization and value addition.

The initiative will also explore avenues for value addition by promoting the processing and commercialization of cottonseed byproducts, thereby diversifying farmers' incomes and improving their livelihoods. Training and business support will target not only production efficiency but also market integration and income stabilization [21].

Enhancing market access through quality and traceability.

A growing segment of consumers and brands—particularly in premium markets—are prioritizing certified organic cotton, transparency, and social standards across the value chain. These actors are willing to pay price premiums, provided product origin and quality are assured. Therefore, quality management becomes a prerequisite for competitiveness [22].

Contamination, improper harvesting, and substandard storage can irreversibly degrade fiber quality and compromise returns across the textile value chain. The project will establish training programmes focused on lint quality assurance, including post-harvest handling, debris control, seed purity, and moisture management. These efforts will be supported by capacity-building in laboratory-based fiber testing, classification systems, and standard compliance. Actors, such as ginners, classifiers, traders, and cotton breeders, will receive specialized training in fiber evaluation and market-relevant classification.

Institutional strengthening and standard development.

In collaboration with the Ministry of Agriculture and Land Reclamation (MALR) and its affiliated bodies, the project will develop an action plan to improve the quality of the Egyptian cotton sector. This includes establishing technical guidelines and national standards covering cultivation, ginning, classification, and trade practices for organic and non-contaminated cotton.

Best practices in traceability and product transparency, as pioneered by Filmar through the NILO organic cotton brand and Italy's Traceability & Fashion certification system, will be scaled and adapted. This model provides consumers with verified information on the origin, environmental performance, and social compliance of products across the supply chain [23].

Innovation and research in cotton varietals.

Egypt's cotton sector is currently constrained by a lack of agronomic research and varietal development. The project will strengthen the capacity of MALR and research institutions to pursue breeding programmes, including seed purification and hybrid development, targeting improved yields and market-aligned fiber traits. Private-sector participation in experimental plantations will be encouraged to reduce stock accumulation and align production with international demand.

Branding, certification, and international positioning [24].

To reverse quality degradation and restore Egypt's reputation in global markets, improvements in seed quality, yield optimization, contamination control, and brand visibility are imperative. The project will support producers and institutions in actively marketing organic and traceable Egyptian cotton to selected high-value markets.

Human capital development and curriculum reform.

Finally, in collaboration with the private sector and educational institutions, the project will co-develop and implement updated training modules, targeting vocational students and technical instructors. These programmes will emphasize labour market alignment, integrating in-company apprenticeships, practical modules, and exchanges to strengthen the skills pipeline in both agriculture and textile industries [25].

9. Conclusion and strategic implications

The imperative of moving up the value chain is clearly demonstrated by the significant increase in value captured at each successive stage of the cotton-textile-garment (CTG) production process. In alignment with Egypt's national priority of reducing reliance on inefficient public-sector enterprises and encouraging private-sector leadership, this project seeks to foster domestic value addition by promoting the local transformation of lint cotton into yarns and finished goods. This shift is expected to generate sustainable employment, enhance income-generation opportunities for local communities, and contribute to more secure and diversified livelihoods [26].

To this end, a comprehensive package of technical, technological, and business support services will be provided to LS and ELS cotton processors. These services will enhance capacity in supply chain optimization, production and quality management, technological upgrading, market responsiveness, and access to finance and partnerships. Leveraging Filmar's existing business network, the project will facilitate market intelligence sharing, buyer engagement, and promotional missions to cotton-importing countries, ensuring that Egyptian producers are well informed of international buyer expectations and aligned with global market trends.

Given that quality is a non-negotiable criterion for trade and market access, the project places strong emphasis on capacity-building in quality management. Targeted training will cover process control in spinning, weaving, dyeing, and finishing, as well as laboratory-based testing and certification of cotton products. Additionally, in line with UNIDO's experience in promoting sustainable production and resource efficiency, the project will introduce RECP methodologies to rationalize the use of water, energy, and chemicals, starting with the systematic metering of resource consumption. These measures will not only reduce environmental externalities but also contribute to operational cost savings [27].

Particular attention will be paid to industrial wastewater management, particularly in dyeing operations, and to innovative circular economy approaches, such as upcycling textile waste. These strategies are crucial for minimizing the industry's environmental footprint and positioning Egyptian cotton as an eco-responsible product in global markets [28].

Recognizing that the sector's competitiveness is also constrained by a shortage of skilled labour and technical expertise, the project will collaborate with TVET institutions, private-sector partners, and policy actors to modernize education and training systems. It will build human capital by equipping young entrepreneurs, managers, and technical workers with a full spectrum of skills, from cotton farming and processing to product design and marketing. The ongoing efforts of Filmar and AlexBank in updating industrial school curricula will be expanded, and local advisory ecosystems will be strengthened to support fashion innovation and SME development.

To overcome systemic constraints faced by individual firms—particularly SMEs—the project will actively promote cluster-based development strategies. Clustering enables economies of scale, knowledge exchange, and shared infrastructure, thereby creating a fertile environment for innovation and competitiveness. Joint initiatives, such as pooled procurement, cooperative marketing, and shared technical services, will enhance the efficiency and visibility of textile enterprises.

Furthermore, the project will explore the feasibility of establishing textile industrial parks or zones, guided by UNIDO's experience in developing industrial ecosystems. Feasibility studies will be conducted in partnership with national stakeholders to assess infrastructure needs and investment potential.

Finally, the project will contribute to the formulation of sector-wide strategies and policy frameworks, co-designed with the Ministry of Trade and Industry (MTI) and affiliated institutions. These will support public-private partnerships in capital-intensive segments, such as spinning, weaving, dyeing, and printing. Capacity-building for MTI and other regulatory entities will focus on improving service delivery to industry stakeholders, promoting innovation, quality assurance, sustainable production, and compliance with international standards.

By integrating value chain upgrading, capacity-building, environmental sustainability, and policy support, the project aims to reposition Egyptian cotton as a competitive, high-quality, and sustainable product in both domestic and global markets.

Author details

Dalia Gallico
Università San Raffaele Roma, Rome, Italy

*Address all correspondence to: dalia.gallico@uniroma5.it

References

[1] Gherzi. Vision 2025 Expansion & Employment – Egypt Textile National Strategy. Gherzi; 2022

[2] Better Cotton Initiative. Better Cotton Production Principles and Criteria Explained. Better Cotton Initiative; 2023. Available from: https://bettercotton.wspreview.net/wp-content/uploads/2023/04/Better-Cotton-PC-v.3.0.pdf [Accessed: April 28, 2025]

[3] Hussein K. Cotton in west and Central Africa: Role in the regional economy & livelihoods and potential to add value. In: Proceedings of the Symposium on Natural Fibres. 2018. Available from: https://www.fao.org/4/i0709e/i0709e05.pdf [Accessed: April 28, 2025]

[4] UNECA, AU. Economic Report on Africa 2023: Egypt Country Case Study. UNECA, AU; 2023. Available from: https://www.uneca.org/economic-report-on-africa-2023 [Accessed: April 28, 2025]

[5] UNIDO. Globalization, the Changed Global Dynamics of the Clothing and Textile Value Chains and the Impact on Sub-Saharan Africa. Research & Statistics Branch, Working Paper 10/2008; 2019b. Available from: https://www.unido.org/sites/default/files/2009-12/Globalization_changed_global_dynamics_of_clothing_and_textile_value_chains_and_impact_on_subSaharan_Africa_01_0.pdf [Accessed: April 28, 2025]

[6] COMESA. Regional Strategy for Developing the Cotton, Textile and Clothing Manufacturing, Marketing and Distribution/Exporting. COMESA; 2019. Available from: https://unctad.org/system/files/non-official-document/Cotton_to_Clothing_Sector_COMESA_1617K.pdf [Accessed: April 28, 2025]

[7] USAID. Improving Productivity in Egypt's Ready-Made Garments Sector. Trip Report. USAID; 2019. Available from: https://studylib.net/doc/8392460/improving-labor-productivity-in-egypt-s-ready [Accessed: April 28, 2025]

[8] Cotton Commodity Briefing. Fair Trade and Cotton. Cotton Commodity Briefing; 2015. Available from: https://www.fairtrade.org.uk/resources-library/commodity-reports/cotton-commodity-briefing/ [Accessed: April 28, 2025]

[9] Gazanfer M. Identification of Strategies for Developing the Cotton Value Chain in West and Central Africa: A Comparative Study on India, Turkey, and Egypt. German–Arab Trade; 2017. Available from: https://downloads.unido.org/ot/47/88/4788840/20001-_23662.pdf [Accessed: April 28, 2025]

[10] GTZ SME Promotion Programme. Textile Desk Research: Value Chain Analysis Preparation Phase. Context Consulting & Services; 2014

[11] UNIDO and Gherzi. Feasibility Study for a Cotton Spinning Mill in 11 Sub-Saharan African Countries. UNIDO and Gherzi; 2021. Available from: https://downloads.unido.org/ot/46/71/4671535/LEE_AGR_AIT_RAF_2008_XPRAF08005.pdf [Accessed: April 28, 2025]

[12] Boons F, Lüdeke-Freund F. Business models for sustainable innovation: State-of-the-art and steps towards a research agenda. Journal of Cleaner Production [e-book]. 2013. DOI: 10.1016/j.jclepro.2012.07.007

[13] UNIDO. Agro-Value Chain Analysis and Development: The UNIDO Approach. Staff Working Paper. UNIDO; 2019a. Available from: https://www.unido.org/sites/default/files/2010-02/Agro_value_chain_analysis_and_development_0.pdf [Accessed: April 28, 2025]

[14] Miska C, Szőcs I, Schiffinger M. Culture's effects on corporate sustainability practices: A multi-domain and multi-level view. Journal of World Business. 2018;**53**(2):263-279. DOI: 10.1016/j.jwb.2017.12.001

[15] Schaltegger S, Lüdeke-Freund F, Hansen EG. Business models for sustainability: A co-evolutionary analysis of sustainable entrepreneurship, innovation, and transformation. Organization & Environment. 2016;**29**(3):264-289. DOI: 10.1177/10860266166332

[16] USDA Foreign Agriculture Services. Egypt Cotton and Products Annual 2016. Gain Report. USDA Foreign Agriculture Services; 2016. Available from: https://www.fas.usda.gov/data/egypt-cotton-and-products-annual-0 [Accessed: April 28, 2025]

[17] Silvestre BS. Sustainable supply chain management in emerging economies: Environmental turbulence, institutional voids and sustainability trajectories. International Journal of Production Economics. 2015;**167**:156-169. DOI: 10.1016/j.ijpe.2015.05.025

[18] Gwilt A, editor. A Practical Guide to Sustainable Fashion. London: Fairchild Books/Bloomsbury Publishing; 2014. DOI: 10.5040/9781350067059

[19] Black S. The Sustainable Fashion Handbook. London: Thames & Hudson; 2012. Available from: https://thamesandhudson.com/sustainable-fashion-handbook-9780500290569 [Accessed: April 28, 2025]

[20] Adams R, Jeanrenaud S, Bessant J, Denyer D, Overy P. Sustainability-oriented innovation: A systematic review. International Journal of Management Reviews. 2016;**18**(2):180-205. DOI: 10.1111/ijmr.12068

[21] Ekins P. Economic Growth and Environmental Sustainability: The Prospects for Green Growth. London: Routledge; 2000. DOI: 10.4324/9780203011751

[22] Bansal P, Roth K. Why companies go green: A model of ecological responsiveness. Academy of Management Journal. 2000;**43**(4):717-736. DOI: 10.2307/1556363

[23] Stiglitz JE, Greenwald BC. Creating a Learning Society: A New Approach to Growth, Development, and Social Progress. New York: Columbia University Press; 2014. DOI: 10.7312/stig15214

[24] Silvestre BS, Țîrcă DM. Innovations for sustainable development: Moving toward a sustainable future. Journal of Cleaner Production. 2019;**208**:325-332. DOI: 10.1016/j.jclepro.2018.09.244

[25] Gaziulusoy I, Brezet H. Design for system innovations and transitions: A conceptual framework integrating insights from sustainability science and theories of system innovations and transitions. Journal of Cleaner Production. 2015. Available from: https://research.tudelft.nl/en/publications/design-for-system-innovations-and-transitions-a-conceptual-framew [Accessed: April 28, 2025]

[26] Sen A. Development as Freedom. Oxford: Oxford University Press; 1999. Available from: https://oxford.co.za/

shop/higher-education/economics-higher-education/9780192893307-development-as-freedom [Accessed: April 28, 2025]

[27] Kassinis GI, Soteriou AC. Greening the service profit chain: The impact of environmental management practices. Production and Operations Management. 2009;**12**(3):386-403. DOI: 10.1111/j.1937-5956.2003.tb00210.x

[28] Fletcher K. Sustainable Fashion and Textiles: Design Journeys. 2nd ed. London: Routledge; 2013. DOI: 10.4324/9781849772778